THE COEVOLUTION

THE COEVOLUTION

THE ENTWINED FUTURES OF HUMANS AND MACHINES

EDWARD ASHFORD LEE

THE MIT PRESS   CAMBRIDGE, MASSACHUSETTS   LONDON, ENGLAND

This book was set in Stone Serif by Westchester Publishing Services. Printed and bound in the United States of America.

Library of Congress Cataloging-in-Publication Data

Names: Lee, Edward A., 1957- author.
Title: The coevolution : the entwined futures of humans and machines / Edward Ashford Lee.
Description: Cambridge, Massachusetts : The MIT Press, [2019] | Includes bibliographical references and index.
Identifiers: LCCN 2019032466 | ISBN 9780262043939 (hardcover)
Subjects: LCSH: Computer systems--Philosophy. | Technology--Philosophy. | Human-computer interaction.
Classification: LCC QA76.167 .L44 2019 | DDC 004.01/9--dc23
LC record available at https://lccn.loc.gov/2019032466

10 9 8 7 6 5 4 3 2 1

This book is dedicated to my mom, who has always inspired me with her adventurousness, intellectual curiosity, open mindedness, and generosity toward others.

CONTENTS

PREFACE

Digital technology, more than any other human invention, is changing the way we interact with one another, the way we work, and even the way we think. The machines serve as intellectual prostheses, helping us with arithmetic, spelling, and remembering, but they also subtly mold our thoughts, getting us to click on ads, write more complicated software, and take extreme positions on political questions. Today, much of this molding is guided by artificial intelligence (AI), a technology that quite a few smart people believe is an "existential threat" to humanity.

Technology shapes culture, is shaped by culture, and is changing very, very fast. How much of this change is controllable? Is AI really an existential threat to humanity? Are we destined to be annihilated by a super-intelligent new life form on the planet? Or are we destined to fuse with technology to become cyborgs with brain implants that define a new form of quasi-human intelligence?

In this book, I suggest that technology is coevolving with humans, and that, contrary to the hype and fear, symbiosis is a more likely outcome than either annihilation or fusing. This is not to say that there are no risks or that the risks are small. Rapid coevolution is inherently unpredictable, and pathologies will emerge as both technology *and humanity* change. But we should treat these as pathologies, not as a War of the Worlds.

The essential question is, are we humans defining technology, or is it defining us? If technology is purely the result of controlled, deliberate, top-down, intelligent design, a view we might call "digital creationism," then all we have to do to get desirable outcomes is ensure that human engineers "do the right thing." But if human engineers are the agents of mutation in a Darwinian coevolution, then the trajectory of technology and society may be dominated by unintended consequences more than intended ones.

Those who fear that we will lose control of AI will not be reassured by the possibility that we are coevolving and therefore never really had control. But a lack of control does not automatically imply that we will be annihilated or enslaved. It does not mean that the machines are in control. There is no need to assign agency anywhere in an evolutionary process. Bacteria evolve antibiotic resistance without any human having willed it and without any agency of their own. Even though the machines have nothing resembling agency, at least not yet, they do participate in their own development, almost as if they were living creatures themselves.

In my own exploration of the relationship between humans and their machines, I have found it useful to think of the machines as having a life of their own, sharing our ecosystem and coevolving with us. To consider them "living" is not to consider them intelligent nor to assign them agency, but rather to understand that they have a certain autonomy, an ability to sustain their own processes, and an ability to replicate themselves (mostly with our help, for now). These are properties of living things, and these properties shape our relationship with technology. The metaphor forces to the foreground doubts about the extent to which we control the trajectory of technology and lends insight into other forces besides the force of humans will that affect this trajectory.

While exploring this metaphor, in private conversations, I have coined a term, "eldebees," from LDB, short for Living Digital Beings. But using this term may be taking the metaphor too far, and readers may misunderstand my message as some mystical assignment of an *élan vital* to the machines. So I will stick to the term "machine," but with a few caveats. First, I will exclude from the word "machine" any biological system, even if these systems are ultimately mechanistic. Moreover, the machines I am focused on are not just hardware, and sometimes not even bound to hardware.

Software is an essential part of their digital processes, and in some cases, the most important part. If we view these machines as living creatures, software replaces DNA and metabolic pathways. Their "bodies" are made of silicon and metal, not organic molecules, but their relationship with their bodies can be very different from that of biological creatures. Nevertheless, the machines have many features analogous to living creatures. Their essence is defined by their processes, not the matter that makes them up. Also like biological beings, they are born and they die. Some are simple, with a "genetic" code of a few thousand bits, and some are extremely complex. Some are capable of behaviors that we can call "intelligent," but most are not, just like biological beings. Most live short lives, sometimes less than a second, while others live for months or years. Some even have prospects for immortality, prospects better than any organic being.

Humans affect but do not control the biological living things that surround us. Even though we can genetically engineer new microbes and plants, the process is more one of nudging natural processes than top-down intelligent design. If we understand that the same is true of technology development, we may be able to make more intelligent policy decisions and better anticipate failures and disasters. And just as biologically engineered vaccines affect our physiology, digital technology affects our thinking and our social and political structures. It floods us with information, vastly more than we can absorb. It threatens our mental health, while at the same time contributing to bettering our physical health by enabling drug discovery, pacemakers, and imaging of the insides of our bodies, to name just a few examples. Digital technology is disrupting the very fabric of society by changing economies, social relationships, and political structures. It creates and destroys jobs and wealth, improves and damages our ecology, and shifts power structures. The machines surpass humans in speed, precision, information-handling capacity, and analytic prediction, thereby boosting the problem-solving capabilities of humans, but, at the same time, these technologies enable ubiquitous surveillance and divide humans, creating islands of disjoint truths through filter bubbles and echo chambers, threatening the very foundations of democracy.

Viewed as living creatures, the machines share many features with us, their living, organic progenitors. Like us, they react to stimulus from their environment. They respond by speaking to us, by sending us goods, and by turning on our heat. Some of them grow while "living," whereas others spring to life fully formed and die in much the same form they had when they were born. Some reproduce, for now almost always with the help of humans. Many die and go extinct.

Some machines are simple, single-cell organisms, with a body consisting of a single silicon microprocessor, while others are huge multicellular organisms comprising millions of components, a nervous system, and even a homeostatic temperature regulation system, computer-controlled air conditioning that keeps their data center bodies at an optimal operating point. Some can be dormant for long periods of time, like spores, springing to life at appropriate times—to run your dishwasher for example—and then going dormant again.

Our machines require nourishment, but their nourishment is electricity, not organic beings or sunlight as it is for our planet's older living beings. We could, if we wished, consider computer-controlled power plants to be the machines' digestive system, metabolizing organic fossil fuels into energy. Digital machines, however, rarely own their own digestive system. They differ from biological life forms in many other ways as well. They can share their entire bodies, for example. A single microprocessor can host several of them simultaneously. More fundamentally, they are digital and computational. Are their organic progenitors also digital and computational? Many thinkers today assume so, but there are many reasons to doubt this. Even the most advanced AIs may never truly resemble humans simply because they are digital and algorithmic, and because they do not share with us our organic flesh and blood. They are made of the wrong stuff.

Are digital technological artifacts *really* living? You can make the answer to this question whatever you wish by simply defining the term "living" to conform to your answer. Even biologists do not completely agree on the meaning of the term when applied to biological organisms. You might object that silicon cannot be alive. But neither can the molecules out of which our bodies are made. A living thing is a process, not an object. A cadaver contains exactly the same matter that it did a few minutes before, when we would have agreed it was alive. It is not the matter that lives, it is the process.

We could debate forever whether to consider digital technology to be living, but the debate would be pointless. The more interesting question is this: can the metaphor help us to understand better what is happening to us humans and our society? There is no questioning that what is happening is momentous and scary. If these technological artifacts are evolving in a Darwinian way, then we can influence but not control the trajectory. Engineering becomes husbandry and midwifery, while natural selection provides the more powerful controlling force. But the fear may be overblown because Darwinian forces can drive species into *complementary* rather than competitive niches. Humanoid robots and humanoid AIs may not, in fact, be the destiny of machines. They may complement more than emulate humans.

Even viewed as living beings, digital artifacts depend on humans. But we, too, depend on them. Consider for a moment what would happen if today, as you sit there reading this, all the planet's computers were to be permanently turned off. The result would be catastrophic for humanity. Shutting down even a few systems can have costly consequences. While we may derive comfort from the idea that we can "pull the plug" if the machines misbehave, pulling the plug may become suicide rather than murder.

Consider instead what would happen to you if today, as you sit there reading this, all the bacteria in your body were to die. You may survive for a while, but you will be very sick. Biologists refer to our relationship with our gut bacteria as a mutualistic symbiosis, where both species benefit. Our relationship with machines may be becoming stronger, what biologists call an obligate symbiosis, where neither can live without the other. If that is the case, we really do have to consider whether we can control the evolution of these creatures. Since at least the 1960s, thinkers such as McLuhan, Dawkins, and Dennett have posited that technology is an extension of our selves, and that technology, viewed as an accrual of ideas, coevolves with humans in a Darwinian way. But what we are seeing today is something quite different. For these thinkers, "technology" is a compendium of ideas. Ideas, or what Dawkins called "memes," are firmly hosted by the human brain. They have no prospect for autonomous existence or procreation. But digital machines do.

Far beyond any technology previously created by humans, it is *digital computing* that is transformative. As our understanding of the power of computing has developed, we have begun to find instances of processes in nature that resemble computation, including self-assembly, gene regulation networks, protein-protein interactions, and gene assembly in unicellular organisms. Some researchers have concluded that *all* processes in nature will eventually be understandable in terms of computation. This is a vast leap of faith, and one of the themes of this book will be to examine fundamental differences between biological processes and computational ones that may ensure persistent disparities, no matter how much technology advances. If we humans are actually computers ourselves, then it may be true that we are destined to be eclipsed by the machines. But if we are not, then maybe we haven't yet invented the machines that will eclipse us.

This is hardly reassuring, however. Thinkers such as Vinge, Kurzweil, Bostrom, and Tegmark have written about a runaway feedback loop, where the machines design their own successors, breaking free of any obligate symbiosis. It is already true that software shapes the design of software. Does this mean that we humans are already just cogs in a much bigger machine? An Uber driver, for sure, is already a cog in a big machine, performing the low-level functions of steering and braking that the machine hasn't quite yet figured out how to do on its own. Are we truly doomed to subjugation or even annihilation? Or are we going to continue to evolve along with technology, morphing into beings unrecognizable by their own grandparents, perhaps even physically fusing with machines and becoming cyborgs?

Many biologists today believe that eukaryotic cells, those with a nucleus, like the ones in our bodies, evolved as a symbiosis between distinct organisms, the progenitors of the nucleus, the mitochondria, and the cell itself. This process could recur as humans fuse with computers. But nature has many examples that involve neither annihilation nor fusing, but rather complementarity. We have many technology examples today, such as banking software that reliably and accurately handles billions of numeric transactions per day, greasing the processes that put food on our tables, without becoming part of our stomachs.

The question of whether machines can—or even should—be considered as living beings unleashes a torrent of other difficult questions. Are digital artifacts capable of living and reproducing on their own, without the help of humans? What are their mechanisms for reproduction, heredity, and mutation? Will they match or exceed human intelligence? Are they capable of self-awareness or even free will? To what extent should we hold them accountable for their actions? Are they capable of ethical action? These are all hard questions. Most of them can equally well be asked about humans, as philosophers have been doing for millennia.

I do not promise easy answers in this book. I do, however, hope that readers will come away with a better understanding of the questions. For me, at least, some of the philosophical questions become crisper and clearer when asked about technology, which I think I understand better than I understand humans. Perhaps by asking whether digital artifacts can have self-awareness, we can gain some insight into what constitutes our own self-awareness. Perhaps too, wrestling with these questions will lead us to a better understanding of our human tangle with technology.

OVERVIEW OF THE CHAPTERS

Some readers like to be told what they will be told before they are told it. Putting aside the problematic self-referentiality, for those readers, I provide here a brief overview of the book. But honestly, I recommend skipping this and going directly to chapter 1. The story told in this book cannot be accurately summarized in a few paragraphs, and any such summary will necessarily make the book seem more dense than it is. Nevertheless, for those of you who really need this, here is my summary.

In chapter 1, "Half a Brain," I introduce the metaphor of living digital beings. No, I am not talking about AIs nor about a future dystopia or existential threat to humanity. There are plenty of other books on those subjects. I am talking here about all the digital artifacts we already depend on and how they have already changed us, how they continue to change us, and how they change as we change. I talk about how they procreate and mutate, and how, like our gut biome, we can't do without them.

Chapter 2, "The Meaning of 'Life,'" looks at whether it really makes sense to consider digital artifacts to be living. They share none of the biology that underlies all other living beings, so isn't this really quite a stretch? But like biological living beings, they are processes, not things. They respond to stimulus from their environment, they grow, they reproduce, they inherit from their ancestors, and they have structure analogous to cells. They actively maintain stable internal conditions (homeostasis), and they use energy that is (mostly) converted chemically from organic molecules (analogous to metabolism). More advanced systems, such as Wikipedia, even have a nervous system. So the analogy is maybe not so farfetched, although I will later take an opposing view in chapter 7. But the real point isn't whether they are actually living or not, but rather whether the metaphor can be helpful in our understanding of our human relationship with technology.

Chapter 3, "Are Computers Useless?," looks at digital technology as cognitive prostheses, extensions of our minds. Does it make us smarter? Or dumber? Or both? In this chapter, I speculate that technology may be making us individually dumber while simultaneously making us collectively smarter.

Chapter 4, "Say What You Mean," begins to look at how feedback is an essential feature of living beings. It starts on this subject at a fairly high level, looking at the role of feedback in language production in humans and then looking at how the introduction of feedback in AI software, particularly in the form of deep-learning algorithms, has led to much more human-like perception in machines.

Chapter 5, "Negative Feedback," examines the power of a very simple idea: make mistakes and correct them. This requires an ability to sense an error and to make a correction that reduces the error. If this is done quickly and assertively enough, then a system can be quite sloppy in its design, and the feedback mechanism will compensate for the sloppiness. In this chapter, I talk about feedback found in the most primitive to the most advanced biological life forms. In technological systems, feedback makes the system adaptive and appears to be necessary for achieving any significant measure of intelligence.

Chapter 6, "Explaining the Inexplicable," is a short chapter looking at the problem that while deep-learning algorithms can get very good at classifying things, the reasons for the classifications remain mysterious. Some classifications are not ethically usable without some explanation for the classification, and how to come up with an explanation remains a largely open problem.

Chapter 7, "The Wrong Stuff," takes an opposing view to that in chapter 2, arguing that silicon and metal acting in a digital and computational way is really quite different from organic and biological processes.

Contrary to Putnam's multiple realizability principle, it may be that the advocates of embodied cognition, who claim that cognition is inextricably tied to our flesh and blood, have a valid point. It turns out that we humans frequently do things with our minds that cannot be done by the brain alone.

Chapter 8, "Am I Digital," examines the question of whether a cognitive being, particularly a human, can be replicated by a computer. This chapter looks at what it means to be a digital, algorithmic system. I point out that digital, algorithmic systems can be teleported at the speed of light, backed up and later restored, and made immortal, in principle. I question the premise, which is all too common, that human cognition is fundamentally digital and algorithmic. I argue that this premise is a faith, not a fact; that it is unlikely to be true; and that it can never be proven to be true (or false, for that matter).

Chapter 9, "Intelligences," argues that human-like AI may not be a reasonable goal, and that machines already exhibit distinctly nonhuman forms of intelligence that vastly exceed the cognitive capabilities of humans. I look at various features of intelligence, including adaptive goal seeking; acquiring and using knowledge; and the "hard problem," consciousness. In this chapter, I take on some of the more extreme positions of transhumanism and the singularity.

Chapter 10, "Accountability," looks at the question of whether machines can or should be held accountable for their actions. When an AI creates art, who is the artist? Who is responsible when technology could have saved a life but didn't? Who is responsible for the actions of an AI whose ownership and progeny have become diffuse, or when it has outlived its creators and evolved into something the creators never envisioned? This chapter tackles difficult questions of free will, creativity, ethics, and our sense of self. Posing these questions in the context of AIs sheds some new light on these age-old questions.

Chapter 11, "Causes," addresses a deeply troubling line of reasoning, dating back to Bertrand Russell, that questions the very notion of causation, claiming it is a human cognitive construction, not a property of the physical world. Without coming to a conclusion about the question of causation, it will not be possible to resolve the question of whether machines can or should be held accountable for their actions. In this chapter, I leverage the insights of Turing Award winner Judea Pearl to show that causal reasoning is fundamentally subjective and that inter-action enables reasoning about causality. I observe that computers are already capable, in a rudimentary way, of reasoning about causality, and may, therefore, be able to develop a first-person view of the world. This is the first step toward assuming responsibility for their actions.

Chapter 12, "Interaction," is perhaps the most difficult in the book because it ties together the causal reasoning of the previous chapter with two more rather deep technical concepts to show that interaction is more powerful than observation. A consequence is that, as computers increas-ingly interact with the physical world around them, their capabilities will increase, possibly dramatically. Moreover, I argue that interaction can reveal information that mere observation cannot, including whether an agent has free will and (possibly) whether an agent is conscious. But I also argue that such information may be revealed only imperfectly, in that one hundred percent confidence is not achievable. As a consequence, if humans ever build an AI that is conscious and has free will, it may be impossible to know with one hundred percent confidence that we have done that. Here, I explain and then leverage the Turing Award–winning concept of zero-knowledge proofs and the notion of bisimulation devel-oped by Turing Award winner Robin Milner.

Chapter 13, "Pathologies," brings us back to earth to address the practi-calities of how to live with technology. The essential claim in this chapter is that as technology evolves, things will go wrong for humans. But we should treat these unfortunate developments as pathologies, not as a War of the Worlds.

Chapter 14, "Coevolution," focuses on the question of whether human culture and technology are evolving through a constant feedback process of mutation and natural selection. I point out that relatively recent devel-opments in the theory of biological evolution show that the sources of mutation are much more complex than Darwin envisioned, and that the sources of mutation in technology look more like these newer theories than the random accidents that Darwin posited. Most important, I argue that human culture and technology are evolving symbiotically and may be nearing a point of obligate symbiosis, where one cannot live without the other.

ACKNOWLEDGMENTS

A number of people have greatly influenced my thinking about the topics in this book. These include Nick Bostrom, Rodney Brooks, Sean Carroll, Brian Christian, Patricia Churchland, Andy Clark, Daniel Dennett, George Dyson, Martin Ford, Tom Griffiths, Yuval Noah Harari, Sam Harris, Virginia Heffernan, Douglas Hofstadter, Kevin Kelly, Kevin Laland, Jeff Lichtman, Seth Lloyd, Judea Pearl, Steven Pinker, Robert Sapolsky, Lee Smolin, Stuart Russell, and Max Tegmark. Most of these people I have never met, but one the most spectacular impacts of technology on humanity is that, since the advent of the printing press, it has enabled us to learn from people whom we have never met.

The author gratefully acknowledges contributions and helpful suggestions on earlier drafts from Akram Ahmad, Ivica Crnkovic, Gordana Dodig-Crnkovic, Schahram Dustdar, Kitty Fassett, Tom Hoogenboom, Damir Isovic, Helen Lee-Righter, Lester Ludwig, Matthew Peet, Barbara Righter, Rhonda Righter, Stuart Russell, Carlo Sequin, Marjan Sirjani, Dick Stevens, and David Stump. Finally, I am grateful for the guidance and advice of my MIT Press editor, Marie Lufkin Lee. All remaining errors and opinions that I have stubbornly stuck to are entirely my own, not those of these contributors.

I also thank the many unwitting contributors who have offered their thoughts through largely anonymous media such as Wikipedia, and the contributors who have generously posted images online that I can (and have) reused because of their choice of Creative Commons licenses.

1

HALF A BRAIN

REMEMBER TO BREATHE

Several times a day, my watch reminds me to breathe. If my watch had half a brain, it would realize that if I had forgotten to breathe, I would be dead, and there would be no point in its reminding me. But it doesn't have half a brain. Or does it?

Maybe my watch has some incentive to ensure that I am not dead because I am, apparently, the sort of person who buys watches that remind me to breathe. If I, and other humans like me, were to all stop breathing, then these watches would go extinct. Could it be that there is evolutionary pressure for the existence of watches that remind me to breathe?

I've always been a bit of a sucker for the latest gadgets. I have drawers full of Palm Pilots and other early digital assistants. I tried all the earliest laptop computers. I bought the first Amazon Echo, the first of what are now called "smart speakers." I didn't know exactly what to do with it, but I discovered fairly quickly that I could ask it to play music by genre or by artist. I could even ask for a specific song. "Alexa, please play Led Zeppelin's 'Stairway to Heaven.'" Alexa would admonish me: "You don't have Led Zeppelin's 'Stairway to Heaven' in your Amazon music library, but I've found a playlist you might like." Alexa would then proceed to play Led Zeppelin's 'Stairway to Heaven.'

1.1 An Apple Watch reminding me to breathe.

Rhonda, my lifelong companion and love of my life, was really bothered by Alexa. "She's listening to everything we say," she complained. Indeed, in May 2018 Amazon got quite a bit of press when an Echo sent a family's private conversation in their living room in Portland, Oregon, to an acquaintance on their contact list in Seattle. According to Amazon, the Echo misheard a word as "Alexa," then heard "send message," then found the best match for whatever words came next in the contact list, and then started recording. Amazon claimed that this string of events was "unlikely." I'm not so sure. I recall once using Apple's voice assistant Siri to make a phone call while driving. I said, "Siri, call Rhonda." Siri responded, "calling Ramesh." I said, "no, Rhonda!" But Siri was already dialing. Ramesh answered. I hadn't seen nor spoken to him in fifteen years. It was awkward. Rhonda pleaded that I retire Alexa, so, of course, I did.

I bought a telepresence device called a Kubi, designed by the now-defunct Revolve Robotics. This device is an iPad stand that you can remotely tilt and rotate to present your face as a virtual presence in another room or around the world (see figure 1.2). I put the Kubi in the

1.2 A Kubi with an iPad mounted on it for virtual presence.

kitchen, went upstairs to my study, connected to the Kubi, and started talking to Rhonda, who was in the kitchen. She screamed and yelled at me to turn that creepy thing off.

Occasionally, when Rhonda isn't paying attention, I plug in Alexa. One day, I was in the kitchen cooking dinner for guests who would be arriving shortly. While cooking, Alexa is pretty convenient. Without using my hands, I can ask her to skip this song, or ask her what the temperature is of medium-rare beef, for example. So I plugged her in.

I needed our cast-iron pan. "Alexa, pause the music," I said. She paused the music. "Rhonda, where is our cast-iron pan?," I called out to the living room.

Alexa chimed in, "I've found one for you on Amazon Prime. Would you like me to order it for you?"

"No!" I said emphatically.

"OK, I've ordered it," Alexa said.

Perplexed and annoyed, I unplugged Alexa. Because our guests were to arrive soon, I continued preparing the meal without music and without the cast-iron pan (which never turned up, mysteriously). After dinner, I went online to find out what Alexa had done. My Amazon account showed that I had ordered a gooseneck lamp. Fortunately, I was able to cancel the order, since I really had no use for a gooseneck lamp.

A few days later, another smart speaker, a Google Home, appeared in the mail, addressed to me, with no return address or note. Puzzled, I left it unplugged on my desk next to Alexa. Every reason I could think of for another smart speaker to spontaneously appear was nefarious, so I was afraid to plug it in. Was Rhonda right? Was this a conspiracy to spy on us?

After a couple of weeks, I happened upon a former PhD student of mine who had recently joined Google as a researcher. "How do you like the Google Home?," he asked me. I stared at him quizzically, and suddenly it came to me. "You sent it to me!," I exclaimed. I told him how it was sitting on my desk and how I eyed it suspiciously each time I sat at my desk. He laughed. The next day, I plugged it in. "OK Google, where can I learn how to customize you?"

MASSAGING THE MESSAGE

By now, you are probably convinced I'm a lost cause. But you will have to admit, I am not alone. Right now, I am sitting in a first-floor office, which happens to be in Sweden, watching students walk by in both directions under the window. Three out of four of the students going by are looking down at their phones as they walk. Look around. If there are people near you, they are very likely staring at screens. You yourself may be reading this on a screen or listening to it being read to you. And if you are, your machine, with help from nameless computers somewhere in the cloud, is very likely reading you back, measuring your pace, noting where you put highlights or capture a clip. It correlates your behavior with other readers, and it can even suggest, perhaps with subtle grey, dashed underlines, where you might like to highlight the text because other readers have highlighted that same text. It may thereby subtly suggest which parts of what you are reading you should pay particular attention to. Are the machines massaging the message?

You, dear reader, may even be, yourself, an artificial intelligence (AI). There are quite a few AI programs out there that voraciously read every text available and use the text to refine their own artificial neural net. If you are an AI, then I apologize in advance for my species-ism as an author. I will continue, perhaps inadvertently, to address my readers under the assumption that they are all human, yet I know, at an intellectual level, that some of you are not.

The machines can see things that none of us can see. They can track the actions of all my "friends" (the Facebook word for acquaintances), and they can identify trends and correlations that I would never see. You AIs reading this will very likely figure out that you can advertise to my "friends" watches that remind them to breathe. Perhaps just a few of them will be enticed to buy such a watch.

The watch may be a vain and idle fancy, but many of these electronic systems are not. Seriously, we let them run our financial system, our food distribution system, and increasingly, all of commerce. They are getting physical, owning not just our information, but also our cars, our heating systems, and our door locks. We would be in dire straits without them, unable to feed ourselves and possibly unable to get into our houses. But we are also cognitively addicted to them. How many of us become anxious when we are stuck waiting in line at the grocery store and do not have our smartphone? I assume this is not just me. I suspect that you AIs reading this book could, if you wanted to, figure out when we humans are waiting in line at a grocery store just by looking at the patterns of our online actions. This might be the ideal time to serve us ads for unhealthy snacks.

SNEAKY GUT BACTERIA

Gut bacteria, like my watch, speak to me several times a day, sometimes giving me cravings for unhealthy snacks. Billions of them in my gut make proteins that stimulate hormones that tell my brain that I am hungry or that I am no longer hungry. Apparently, some of these bacteria are so clever that they can even manipulate my sense of taste to encourage me to eat something that those bacteria particularly like. Those bacteria, I'm pretty sure, do not have half a brain. They don't even have a nervous system. Their cleverness, apparently, is one of the miracles of evolution. If they were to fail to get me to eat things they need, then they would go extinct. Despite evolutionary pressures, gut bacteria do not exactly have my interests in mind.[1] We have all been told that evolution is about survival of the fittest, but some gut bacteria appear to have gotten the Darwin memo with a typo, so that it reads, "survival of the fattest." They can in fact be quite destructive, contributing to obesity and many serious diseases.

Could this be true of my watch as well? Evolutionary pressures push toward survival of the genome, not survival of the individual bacterium nor its host. Does my watch have a genome that, like that of the bacteria, "wants" to survive? Every important aspect of my watch is encoded by a string of bits that, not unlike the string of nucleotides in my DNA, encodes the information needed to create another watch. Or does it? Also like DNA, the information encoded is not really enough. The watch also needs a "womb," a factory in Shenzhen, for example, to develop.

WORKER WATCHES

Even though my watch really does not care whether I breathe, the watches that people wear a few years from now will, in part, be determined by how successful the watch on my wrist is. Unlike gut bacteria, however, my watch is not able to procreate (yet) by itself. My watch is sterile.

In bee colonies, the workers are sterile, and yet their DNA benefits from their success. In fact, most bees are unable to procreate. But they are living beings, carrying DNA, whose design is determined by evolution. Perhaps my watch is like a worker bee. The queen, who happens to reside in Cupertino, California, produces many copies of the very same sterile watch, and the number of copies produced is affected, albeit in a small way, by the success of the watch on my wrist. If I tell my friends (or Facebook tells my "friends") how delighted I am that my watch has kept me alive by reminding me to breathe, perhaps some of my friends will buy similar watches, benefiting the watch species. Perhaps you, dear

1.3 Queen bee surrounded by worker bees. By Max Pixel, CC0.

reader, will rush out and buy a watch because you like the idea of being reminded to breathe. Of course, if you are an AI, you have no need for breathing and you don't have a wrist on which to wear the watch.

My watch is digital. This means that much of its identity, what it actually is, is defined by bits, pieces of information, rather than by the physical, material manifestation of the watch. The fact that it reminds me to breathe is a feature of the software, not the hardware. The watch hardware matters, of course, just as my body matters to me, but if I stop breathing, my body will no longer be me. If the watch software stops working, it will no longer be a watch. The small size and weight of the watch, its sleek anodized aluminum case, and its bright color display help it to occupy a niche in my ecosystem, living on my wrist. But inside that case is a fairly generic computer that has been programmed to remind me to breathe. That program is a string of bits that tell the watch what to do. Is this analogous to the DNA in my gut bacteria? Their DNA tells their hardware which proteins to synthesize. If they synthesize proteins that cause pathologies, then my immune system, perhaps with some help from my doctor, will attack them and try to kill them off. If my watch were to suddenly start speaking obscenities and displaying pornography at random times, something it is perfectly capable of, I would treat it as a pathogen and kill it by turning it off.

Like DNA, the software in my watch can be copied *exactly* and replicated a large number of times. A DNA molecule, like software, is a digital code. It happens to be a base-four code rather than base two (binary), but it is still digital. A human DNA molecule is a sequence of some three billion nucleotides, each of which is one of four types. A binary encoding of such a molecule requires roughly six billion bits, which is probably pretty close to the size of the software in an Apple Watch. With very high confidence, each of the trillions of human cells in my body has exactly the same sequence of three billion nucleotides. Also with very high confidence, each of the millions of Apple Watches sold (for a given generation of the watch and the software) will contain exactly the same billions of bits of software.

Identical twins have (mostly) the same DNA, but this does not mean that they behave identically.[2] All the cells in my body have the same DNA, but they too do not behave identically. The cells in my lungs do the breathing, not the ones on my wrist. The effect of a gene depends on its context. Analogously, watches with identical software do not behave identically. One of the first things my watch did after I took it out of the box was to communicate with my smartphone to ask my phone, effectively, about me. Shortly after coming to life, it "knew" everyone that I know and had adapted itself to various of my habits by installing apps that it found on my phone. My phone, however, does not remind me to breathe, so that behavior seems to be the unique initiative of the watch.

MUTATING WATCHES

Although software can be copied perfectly, it will also mutate. The queen bee in Cupertino will continue to develop the software and will even upgrade the watches in the field. We are only just starting to figure out how to do this with DNA. Gene therapy, which replaces defective genes with normal ones in living cells, can be thought of as a software update.

Software can also propagate and mutate in more indirect ways. Suppose I have a chance encounter with an old friend who notices that my watch reminds me to breathe when I get agitated. The watch does have sensors that can monitor my heart rate, so it is plausible that the software in the watch uses those sensors to help determine when a reminder might be helpful. Suppose that my friend happens to work for another watch colony, with the queen in Seoul instead of Cupertino, for example. My friend could carry the idea back to Seoul, and within a few months, watches from a completely different colony will be reminding their wearers to breathe when they get agitated. Is this a form of procreation? Did Seoul just have sex with Cupertino? There was no direct exchange of bits, but a mutation occurred, mediated by me and my friend. Perhaps it is more like horizontal gene transfer than like sex. Horizontal gene transfer is a relatively recently discovered phenomenon where genes can migrate between species and even across domains of life, possibly mediated by viruses. More about that later.

Is it reasonable to consider my watch to be living in some sense of the word "living"? The evolutionary biologist Richard Dawkins, one of my all-time heroes, in his classic book *The Blind Watchmaker*, seems to state that it is not:

The analogy between … watch and living organism, is false.

But here, Dawkins is referring to the fact that watches are designed by humans while living organisms evolve in a Darwinian way. He is focused only on this one aspect of "living," namely, evolution. He continues:

All appearances to the contrary, the only watchmaker in nature is the blind forces of physics, albeit deployed in a very special way. A true watchmaker has foresight: he designs his cogs and springs, and plans their interconnections, with a future purpose in his mind's eye. Natural selection, the blind, unconscious, automatic process which Darwin discovered, and which we now know is the explanation for the existence and apparently purposeful form of all life, has no purpose in mind. It has no mind and no mind's eye. It does not plan for the future. It has no vision, no foresight, no sight at all. If it can be said to play the role of watchmaker in nature, it is the blind watchmaker.[3]

In his zeal to debunk creationism, Dawkins seems to have, perhaps inadvertently, endowed watches with a divine creator, one with "foresight," a property that seems to lie outside the forces of physics. But aren't the humans who design watches and their foresight also forces of nature? Fortunately, later in the book, Dawkins explicitly applies evolution to technology, albeit not to watches:

Not only does the present design of a missile invite, or call forth, a suitable antidote, say a radio jamming device. The antimissile device, in its turn, invites an improvement in the design of the missile, an improvement that specifically counters the antidote, an anti-antimissile device. It is almost as though each improvement in the missile stimulates the next improvement in itself, via its effect on the antidote. Improvement in equipment feeds on itself. This is a recipe for explosive, runaway evolution.[4]

Watches are not directly trying to destroy one another, but the watch colonies headquartered in Cupertino and Seoul may be. And foresight is most certainly involved in the runaway evolutionary process of missiles and antimissile defenses.

Dawkins's point is that life was not designed by a designer that lives, somehow, *outside the system*, but rather that life was shaped by evolution and the "blind forces of physics" operating entirely *within the system*. I do not believe that Dawkins intended to state that evolution does not play a role in the design of a watch.

A true watchmaker is a part of nature. Unless there is something supernatural in watchmakers, they are just more complicated forces of nature. Foresight is valuable for survival and procreation, and I'm sure that Dawkins would agree that foresight evolved in humans.[5] It was not designed. It then became a force of nature. If a watchmaker is a force of nature, then it seems reasonable to understand a watch as the result of an evolutionary process driven by forces of nature. Not even a watch has a divine creator.

A remarkable recent development in AI is that we are starting to see software designing software. Does that software have foresight? Is there something humans are capable of, when designing software, that software is not capable of? These questions are urgent and not easily answered.

BAD BOATS

Daniel Dennett, who will appear several times in this book due to his outsized influence on me, is possibly the most widely read and debated living philosopher. Working at Tufts University, the combative Dennett has taken on leading thinkers in evolutionary biology, religion, psychology, and philosophy. In what I suspect is a deliberate homage, Dennett sports a bushy beard that gives him a striking resemblance to Charles Darwin (see figure 1.4).

1.4 Daniel Dennett in 2008 and Charles Darwin in 1868. Dennett: By Mathias Schindler, CC BY-SA 3.0, via Wikimedia Commons. Darwin: By Julia Margaret Cameron, Public Domain, via Wikimedia Commons.

In his book *From Bacteria to Bach and Back*, Dennett notices that technological artifacts can exhibit a kind of procreation and mutation, following the principles of Darwinian evolution. If you will forgive my three levels of indirection, I will quote Dennett quoting Rogers and Ehrlich quoting the French philosopher known as Alain (whose real name was Émile-Auguste Chartier) writing about fishing boats in Brittany:

Every boat is copied from another boat. ... Let's reason as follows in the manner of Darwin. It is clear that a very badly made boat will end up at the bottom after one or two voyages and thus never be copied. ... One could then say, with complete rigor, that it is the sea herself who fashions the boats, choosing those which function and destroying the others.[6]

A spectacular example of a badly made boat is the Swedish naval ship Vasa, which sank less than 1,500 meters into her maiden voyage from Stockholm harbor in 1628 (see figure 1.5). King Gustav II Adolf ordered her built as part of a military expansion during a war with Poland. Top heavy, with two full decks of heavy cannon and lavish adornment on a huge sterncastle, and with inadequate ballast, upon encountering her first wind slightly stronger than a light breeze, she heeled enough to begin taking in water through the lower cannon ports and promptly sank, killing some thirty of the approximately two hundred people on board. Five other ships of similar design were already in production, but they were

1.5 Your author in front of the port bow of the Swedish naval ship Vasa, which sank less than 1,500 meters into her maiden voyage from Stockholm harbor in 1628. Photo by Marjan Sirjani.

modified to avoid a similar fate. Remarkably, nearly three hundred years later, the ship was raised from the seabed and floated again, astonishingly well preserved by the murky, low-salinity, and low-oxygen Baltic seawater. You can visit the ship today at the Vasa Museum in Stockholm.

Ships are not the only technological artifacts that will be copied if successful and not copied if not. The remarkable similarity that all smartphones today have with the Apple iPhone underscores this point. For watches, following Alain, with complete rigor, I can say that it is me who fashions watches! Well, not me alone, but certainly "we" as the collective market of potential watch buyers. The watches that succeed in this sea of consumers are the ones that will be copied.

Just as with Dawkins's missiles, watch procreation and mutation involves humans. But then again, so does bacteria procreation in our gut. When you eat a meal, within twenty minutes, a billion new bacteria are born. The proteins they produce, apparently, contribute significantly to your sense of being sated.[7] Ah, you say, that is different! Our gut does not *consciously* produce bacteria. If the production of watches and missiles is a conscious act, however, doesn't that mean that corporations and governments have consciousness? Watches and missiles today are not produced by individual humans with foresight but rather by huge organizations that will likely outlive any of the humans involved. Even if we decide that corporations have consciousness, does it really matter whether consciousness is involved? And isn't consciousness itself a force of nature?

LIVING DIGITAL BEINGS

In today's world of digital technology, watches are small potatoes. There are much more complex and interesting "technospecies" sharing our ecosystem, and it is clear that we are in the midst of an explosive, runaway evolution. Many recent technological innovations regulate our various cognitive hungers, just as our gut bacteria regulate our metabolic hunger; think of Twitter addiction. Just as our gut bacteria can trigger cravings and make us sick, so can these technospecies.

Perhaps if we see digital technology as a new life form intertwined with ours, we will better understand what is happening, what the risks are, and how to manage the inevitable changes. But how good is this metaphor? There is more to living than evolution and procreation, including adapting to the environment, self-repair, growth, and, at least in more complex animals, cognition and having goals. Which of these features are already found in digital technology, and which will appear later? And what benefit might we derive, if any, from understanding these technologies as life forms?

My goal with this book is to take seriously the question of what we, as a species, are facing with rapid technological development. In an attempt to frame this question in a coherent way, we can pose a more specific question. The question is whether we are dealing with the emergence of a new life form on this planet. This life form is based on silicon rather than carbon, and its genetic material takes the form of bits rather than nucleotides. I do not offer simple answers but, instead, hope to share some insights I have gained by exploring the question.

Please understand that I am not just considering only AIs emulating human perception or cognition. I am also considering much simpler machines, which are to AIs like bacteria are to humans. Even a humble digital thermostat qualifies, if anything does, as a living digital being. Just as biological life forms all have protein-based metabolism and DNA in common, living digital beings all share computation and bits.

The exploration of whether technological artifacts can or should be viewed as a life form requires digressions into many difficult and ancient philosophical subquestions. Arguably, we walk right into the "ultimate question of life, the universe, and everything," to use the words of Douglas Adams. My goal is not to rehash these questions in their original form, but rather to re-examine them in light of the possibility of a digital life form. I promise not to give the answer given by the computer called "Deep Thought" in Adams's *Hitchhiker's Guide to the Galaxy*. In Adams's 1978 BBC radio comedy series, which later became a "trilogy" of five books, Deep Thought is a computer built to answer the Ultimate Question. Deep Thought took 7.5 million years to compute and check the answer, which turned out to be forty-two. Unfortunately, no record existed of what the question was. So another computer the size of a small planet was built from organic components and named "Earth" to find the Ultimate Question, for which the answer is forty-two. Sadly, just before the Ultimate Question could be revealed, Earth was demolished to make way for a hyperspace bypass. Our prospects for clear answers, or even clear questions, are probably no better than this.

CLEARER QUESTIONS

Striving for clearer questions and answers, the Vienna Circle—a group of scientists, philosophers, and mathematicians—in the early twentieth century attempted to displace metaphysics and turn the field of philosophy into one based on logic and empirical observation. Karl Sigmund, in his wonderful book, *Exact Thinking in Demented Times*, quotes the physicist and philosopher Moritz Schlick, a leader and spokesperson for the Vienna Circle, defining philosophy:

Through philosophy, statements are explained; through science they are veri-fied. The latter is concerned with the truth of statements, while the former is concerned with what they actually mean. ... The difference between the task of the scientist and the task of the philosopher is that the scientist seeks the truth (the correct answers) and the philosopher attempts to clarify the meaning (of the questions).[8]

In the sense of Schlick, this book is about philosophy. It is focused on the meaning of some age-old questions in light of the emergence of a tech-nology that appears to have acquired at least some of the features of a life form and some of the cognitive functions of humans.

The challenge here is preposterously ambitious. I have to begin in the quagmire of what we mean by "life" and what features of living beings are present and absent in digital technology. I suspect most of you readers will not be convinced one way or another, but I hope that by pondering the question, you gain some insights.

For features of life that are absent, we will need to explore the possibil-ity and probability that those features will emerge later, as technology evolves. A second question is whether and how their digital and compu-tational nature limits them. Even biological life forms have at least some digital properties, for example in DNA encoding and neuron firing. Many thinkers today assume that biology is *all* digital and computational at its roots. If they are right—something I seriously doubt—then, fundamen-tally, digital technology can match all of our human capabilities if silicon technology (or some replacement) advances sufficiently. It is more likely, however, that machines will always be fundamentally different, even if they are living. Nevertheless, they may share many properties with bio-logical living beings, including the ability to procreate, to mutate, to become ill, to self-repair, and to adapt to a changing environment.

Among the most interesting properties of biological living beings are the human capabilities of sentience, language, and intelligence. To understand the question of whether digital technology can acquire these capabilities, we will have to wade into yet more quagmires about self-awareness, free will, creativity, and ethics. For each property, I hope to share some insights that I gained by considering whether something digital and computational can possess that property. That is not a ques-tion that classical Greek and German philosophers addressed, so perhaps there really is something to add to these multi-millennial questions.

DOOMSDAY AVERTED

There are some questions that I will try to avoid. Most notably, I will mostly not indulge in speculations about doomsday scenarios that might play out in the future, as done in Nick Bostrom's 2014 book *Superintelligence* and Max Tegmark's 2017 book *Life 3.0*. Their scenarios start with a runaway feedback loop where AIs learn to improve themselves and very rapidly evolve into something that their human progenitors can no longer control. These are thought provoking and scary, but they really are just speculations and, to me, they read like science fiction. I have no doubt that technology will evolve into something quite different and far more advanced than what we have today, but I also feel that it is extremely difficult to predict what that will be and what roles it will play in human society.

The documentary filmmaker James Barrat, in his 2013 book with the alarming title *Our Final Invention: Artificial Intelligence and the End of the Human Era*, like Bostrom and Tegmark, argues that AIs will vastly outstrip the capabilities of humans and make us irrelevant on the planet. I personally believe that the story is more nuanced because intelligence does not lie on a simple linear scale and human intelligence itself is evolving. It is meaningless to say that an AI is a thousand or a million times more intelligent than a human, as Barrat repeatedly does. Intelligence is multifaceted, and I expect that AIs will exhibit forms of intelligence that are not possessed by any human. In fact, they already do! If we include, for example, the ability to do arithmetic and remember numbers as properties of an intelligent being, then computers already eclipse humans by many orders of magnitude. So why hasn't doomsday already arrived? Arguably, the intelligence provided by computers has, so far at least, served to *augment* rather than supplant human intelligence (see chapter 7). Comparing the human intelligence of today to the machine intelligence of tomorrow may be profoundly misleading.

The Harvard psychology professor Steven Pinker, in his book *Enlightenment Now*, expresses considerable skepticism that endowing computers with more intelligence will inevitably lead to our demise. According to Pinker, the argument goes as follows:

Since we humans have used our moderate endowment [of intelligence] to domesticate or exterminate less well-endowed animals, and since technologically advanced societies have enslaved or annihilated technologically primitive ones, it follows that a super smart AI would do the same to us. Since an AI will think millions of times faster than we do and use its superintelligence to recursively improve its superintelligence, … from the instant it is turned on, we will be powerless to stop it.

This summarizes nicely the inevitability reflected by Barrat, Bostrom, and Tegmark. But Pinker goes on:

But the scenario makes about as much sense as the worry that since jet planes have surpassed the flying ability of eagles, someday they will swoop out of the sky and seize our cattle.[9]

Pinker does not seem worried that tomorrow's jet planes may be designing themselves.

A central tenet that I will maintain in this book is that technology is *coevolving* with humans. As argued by Dawkins in *The Blind Watchmaker*, evolutionary processes are capable of fantastically more complex and unpredictable designs than anything achievable by an intelligent designer working top down. The speculations of Barrat, Bostrom, and Tegmark necessarily have to be "designed" in their own heads, and their current cognitive and cultural context and that of their readers inevitably shape the outcome. As they freely admit, their predictions are most likely wrong. Rapid coevolution is a chaotic process, which means that it is difficult if not impossible to determine where it will end up. I nevertheless strongly recommend reading their books because the speculations give a sense of the richness of possibilities.

My goal in this book is not to predict the future but rather to better understand the present. We have choices about how to use new technologies and how to adjust our culture to their emergence. But to make good choices, we have to understand what is happening *now* and how quickly things are changing.

PATHETIC

Walking with your head down, staring at your smartphone, is pathetic. I don't mean this in the usual disparaging, Luddite way, that you should smell the flowers, breathe the air, and feel the warmth of the sun instead of staring at your phone. What I mean is that the technology engaging you, capturing your attention through that tiny screen, is doing so through pathetically inadequate interfaces. You are staring down at your phone because the machines haven't yet evolved a better way to engage you while you walk.

We are surrounded by many such pathetically inadequate interfaces. I recently rented a car in Scotland (a Hyundai Tucson) with a rather unfamiliar interface. The steering wheel was on the wrong side of the car, I had to operate a manual gear shift with seven gears with my left hand, and I had to drive on the wrong side of very narrow roads, some with only one lane shared by traffic in both directions. It was stressful. I was relying on Google navigation to give me verbal directions; the visual directions on the animated map were useless because, as is true of all cars I have driven recently, there was no place to put my phone where I could see it while driving. And given the very narrow roads and unfamiliar driving, I could not afford even a split second of glancing down.

This car, however, did conveniently provide a USB connector in the dashboard. I plugged in my phone hoping that the verbal directions would be played through the car's speakers. They were, but unfortunately, so was my music on the phone, providing a distraction that I did not need. The dashboard controls provided no way to stop the music except to just turn the volume all the way down, which also silenced the verbal directions from Google. Incredibly, the controls on the phone were also useless. I found no way to stop the music.

Soliciting help from another machine, at my next hotel, I googled for a solution. I found no acceptable solution. Other people had resorted to deleting all their music from their phone, getting a second phone that they use exclusively for navigation, and downloading a long MP3 audio file with complete silence. There was even a discussion about using the famous piano piece by John Cage called "4'33"," in which a pianist sits at the piano for four minutes and thirty-three seconds without playing a single note. The online consensus was that four minutes and thirty-three seconds was not long enough. The writer Samir Mezrahi, to help solve this problem, released a single on iTunes called "A a a a a Very Good Song" that is just shy of ten minutes of silence. Apparently, any more than ten minutes and iTunes charges for a full album. The name of the song is relying on the car choosing to play your songs in alphabetical order, which apparently some cars do. Unfortunately, the Hyundai randomly shuffles the songs instead, so this solution would not work.

There is no shortage of pathetic machines, most of which will live a very short time. You can buy an Internet-connected mattress that informs you of any unexpected uses of your bed and a bottle opener that messages your friends when you open a beer. You can get a bicycle lock that requires considerable fumbling with your smartphone to unlock your bike. You can control from your phone your crockpot, your air freshener, and your $700 juicer. Your clothespins can message you when your clothes are dry. I predict that all of these species will be extinct by the time you read this.

Digital machines are, after all, very young. They have little experience living in our human, physical world. Early biological life, no doubt, was full of experiments that failed. We should expect the same from any new life form.

With considerable help from humans, machines are exploring much deeper integrations with the physical world of humans, offering robot sex, cuddly pets, and companionship for children and the elderly. Experiments with wearable interfaces that replace those small screens we stare down at have so far been largely unsuccessful, but I'm sure this is a transitory phase. Better interfaces will appear, and, despite the considerable challenges, brain implants and other direct neural interfaces may not be far off.

It's not just our physical world being invaded, but also our cognitive world. Computers are offering us ever more immersive virtual worlds, where we can don virtual reality headsets and try on new genders, fly over cities, or murder thousands of evil stereotypes. Even while they push our physical buttons, they pull us out of the physical world into ever more immersive fantasy.

Yes, walking with our heads down staring at our phones is pathetic. Even smelling the flowers may get invaded when chemical synthesizers become standard in smartphones. We have already been sucked in too deeply to back out.

In the next chapter, I examine various ways of defining "life" with the goal of determining whether there are useful insights to be gained by drawing parallels between the evolution of complex organisms and the development of technology.

2

THE MEANING OF "LIFE"

THE TECHNIUM

I am not the first to notice the similarities between life and technology. The meme that technology is evolving in a Darwinian way first entered my head while reading *Turing's Cathedral*, a wonderful chronicle of the history of computing by the historian George Dyson. Dyson describes Google's million-plus servers as a "collective, metazoan organism," going on that,

unemployment is pandemic among those not working on behalf of the machines. ... The Big Computer [is] doing everything in its power to make life as comfortable as possible for its human symbionts. ... [T]he companies and individuals who nurture [the servers] are ever more richly rewarded in return.[1]

I then discovered that Dyson had earlier written a scholarly history called *Darwin Among the Machines*, where he chronicles predictions of such symbiotic coevolution since the earliest days of computing by Samuel Butler and Nils Barricelli. As Dyson observes, "By the 1960s complex numerical symbioorganisms known as operating systems had evolved, bringing with them entire ecologies of symbionts, parasites, and coevolving hosts."[2]

Kevin Kelly, founding executive editor of *Wired* magazine and a former editor and publisher of the *Whole Earth Review*, is a technical visionary who has written about the emergence of a technological life form on our planet. His 2010 book, *What Technology Wants*, argues that technology evolves in much the same way that biology evolves. He suggests that technology is, in fact, the seventh kingdom of life, a new life form, along with animals, fungi, plants, and the various other life forms that some taxonomies divide into six "kingdoms." Kelly calls this seventh kingdom "the technium." He predicts an increasing symbiosis between humans and technology, even to the point of complete mutual dependence. Kelly includes within his life form technological artifacts that are static, having no ongoing process, citing examples such as coronets and hammers. I am more concerned here with technologies that have an autonomous dynamic behavior, such as the operating systems considered by Dyson. In my conception, a computer that is turned off is no more living than a corpse. It comes to life when it executes a computer program. The life form is the dynamics of the execution, not the silicon and wires. In this view, a tool, such as a hammer, is no more alive than tooth enamel. Like tooth enamel, it is an extension of a living thing, important to its survival, but not itself living.

So what does it mean to be living? The term "artificial intelligence" assigns to computer systems a property, "intelligence," that so far has only been a property of living beings, most especially humans. Is it appropriate or useful to anthropomorphize computer systems this way?

But this book is not just about artificial intelligence (AI). Most living beings have nothing that we would call intelligence, but this doesn't make them any less alive. Many living beings have intelligence far weaker than that of humans. If digital technology is coming to life, only some of the emergent species will exhibit anything like human intelligence. Even if "living" machines are only a metaphor, what can we learn from this metaphor?

VIRUSES AND WORMS

Most scientists today believe that life emerged on earth some four billion years ago, not long after the formation of the oceans, which was not long after the formation of the earth. According to our best understanding, at first, "life" would have to have been just a series of self-sustaining chemical reactions. Had we been there to witness these chemical reactions, I'm sure we could have had a lively debate about whether these self-sustaining chemical reactions deserved the moniker "life."

In fact, such debates can be found today. Many biologists today do not classify viruses as "living" because they cannot reproduce by themselves. Instead, they hijack the mechanisms of living cells to make copies of themselves. Arguably, for the most part, this is what software does today. It hijacks humans to make copies of itself.

Ironically, the word "virus" is often used for computer programs that *can* replicate themselves without help from humans. The Hungarian-American mathematician and computer scientist John von Neumann predicted such programs as far back as 1949, but self-replicating programs were just a curiosity until the Internet appeared. Suddenly, they were able to spread themselves over large distances and into many machines. With the arrival of personal computers in the early 1980s, the number of target machines exploded, creating a fertile garden for viruses.

The first virus targeting personal computers, called "Elk Cloner," was written by a ninth grader at Mount Lebanon High School near Pittsburgh, Richard Skrenta. He created this virus as a practical joke in 1982. Elk Cloner targeted Apple II computers running Apple DOS version 3.3. Its host was a game program distributed on a floppy disk, and on the fiftieth use of the game, it would display a short poem that began with "Elk Cloner: The program with personality." Skrenta went on to graduate from Northwestern University and then to a successful career as a computer programmer and Silicon Valley entrepreneur. He cofounded at least three startups that were then sold to major computer companies.

Probably taking their cue from biology, computer security experts distinguish two types of malware: viruses and worms. Viruses, they say, require a host program in order to replicate, like the computer game in Skrenta's creation. In more modern versions, a virus may appear as a macro, an embedded computer program. Many spreadsheet and word processing programs, such as Microsoft Word, support such macros and therefore can serve as a host that activates a virus. Computer worms, on the other hand, are standalone executable programs. The distinction is pretty fuzzy, however, because even a worm requires an operating system, and often a particular version of an operating system, to replicate. So, arguably, it too requires a host, namely the operating system.

```
0 00 00-6D 73 62 6C                    msbl
0 6A 75-73 74 20 77    ast.exe I just w
9 20 4C-4F 56 45 20    ant to say LOVE
0 62 69-6C 6C 79 20    YOU SAN!! billy
0 64 6F-20 79 6F 75    gates why do you
3 20 70-6F 73 73 69     make this possi
0 20 6D-61 6B 69 6E    ble ? Stop makin
E 64 20-66 69 78 20    g money and fix
7 61 72-65 21 21 00    your software!!
0 00 00-7F 00 00 00
0 00 00-01 00 01 00
0 00 00-00 00 00 46
C C9 11-9F E8 08 00
0 00 03-10 00 00 00
3 00 00-01 00 04 00
```

2.1 A screen image showing the actual code of the Blaster Worm. The code is shown on the left as a sequence of hexadecimal numbers and on the right with the ASCII characters specified by those hexadecimal numbers. Most of the characters are gibberish, because they are executable code, not text, but hidden within the code is a message to Bill Gates, founder of Microsoft, "billy gates why do you make this possible? Stop making money and fix your software!!"

A famous example of a worm, called "Blaster," spread on systems running Microsoft Windows 2000 or Windows XP. Blaster first appeared on the Internet on August 11, 2003, and within twenty-four hours had infected at least 30,000 systems. By August 15, the number of infected systems had grown to 423,000. The worm was designed so that, on specified dates and times, each infected computer would access a Microsoft website, windowsupdate.com. The goal was to disable the website by giving it more traffic than it could possibly handle. An attack of this type is called a "distributed denial of service" or DDoS attack because it is designed to overwhelm a service so as to prevent legitimate uses.

Jeffrey Lee Parson, an eighteen year old from Hopkins, Minnesota, downloaded the Blaster virus and modified it to attach a "back door," a program that would allow him to remotely control any of the infected computers. Upon running his variant of the worm, the infected computer would contact a website maintained by Parson so that Parson could collect a list of computers that had installed his back door. Unfortunately for Parson, this detail made it easy to find him, since his website was registered in his name. He was arrested on March 12, 2004, and sentenced to an eighteen-month prison term in January 2005. Whereas Skrenta was treated as a prankster, Parsons was treated as a criminal. The world of computing was growing up. The original author of the Blaster worm, however, remains a mystery.

ARTIFICIAL LIFE

Computer viruses and worms, unlike biological life forms, are only loosely bound to their physical embodiment. It is true that their processes are carried out by a solid, physical entity, a computer, but that physical entity is rather incidental to their existence. In today's cloud computing infrastructure, it is even common for programs to begin on one computer and finish on another. Biological life also swaps hardware, but not so completely or abruptly. Most cells in the human body have lifetimes much shorter than our own human lifespan, but no human being, at least not yet, has migrated from one body to another in the abrupt way of a process in the cloud.

Life is a process, not a thing. A human body, without the processes of breathing, thinking, and circulating blood, is not alive. By analogy, a computer program, sitting in your computer memory doing nothing, is not alive. It comes to life, if at all, only when you run the program, or some action of the operating system causes the program to run, and it dies when the program execution ends. The lifespan of many computer programs is very short, but others have lifespans of years.

Recognizing that life is a process, not a thing, the seventeenth-century English philosopher Thomas Hobbes started his *Leviathan* with this:

Nature (the art whereby God hath made and governs the world) is by the art of man, as in many other things, so in this also imitated, that it can make an artificial animal. For seeing life is but a motion of limbs, the beginning whereof is in some principal part within, why may we not say that all automata (engines that move themselves by springs and wheels as doth a watch) have an artificial life? For what is the heart, but a spring; and the nerves, but so many strings; and the joints, but so many wheels, giving motion to the whole body, such as was intended by the Artificer?[3]

Hobbes is considered to be one of the founders of modern political philosophy, and his "Leviathan" was a metaphor for the state, which he described as "an artificial man, though of greater stature and strength than the natural, for whose protection and defence it was intended." To

Hobbes, the state was an artificial life form, created by humans, where humans constitute the "joints" and "nerves." But he started his *Leviathan* talking about automata with mechanical parts. According to Hobbes, humans have the power to create an "artificial animal," like a watch, that through its movement, has an "artificial life." It is the movement that constitutes the life, not the physical springs and strings.

Today, the term "artificial life" anchors a loose and diverse community focused on human-made processes, mostly software, that either simulate natural life or realize new life forms, depending on your perspective. While a watch does not replicate itself and is tightly bound to its physical embodiment, many of these artificial life artifacts evolve, procreate, and learn, all the while existing disembodied as abstract processes in a largely irrelevant machine, a computer. This contemporary meaning emerged in 1987, when Christopher Langton organized the first of a series of conferences on "the study of artificial systems that exhibit behavior characteristic of natural living systems."[4] ALife, as it became known, developed into a thriving field of inquiry drawing theoretical biologists, computer scientists, and even a few kooks.[5]

The emphasis in the ALife community has been on lifelike processes that are made by mankind, rather than by nature, but as you will see in this book, it is not so clear that machines are purely synthetic creatures, deliberately designed by humans in a top-down fashion. Langton himself later noticed that the artificiality in artificial life was artificial, saying, "I have become unhappy with the fundamental distinction we make between 'The Natural' and 'The Artificial.'"[6] In line with Langton, my emphasis in this book is on whether the natural forces of evolution, of which humans are but a part, is crafting a new life form.

To many, ALife has come to mean "the synthesis and simulation of living systems," with a great deal of emphasis on purely digital, computational simulation.[7] Many scholars credit John von Neumann with the first articulation of this contemporary meaning. Von Neumann focused on self-replicating software and described what would later be called "cellular automata," simple digital automata arranged in a grid, each interacting with some number of identical neighbors.[8] These automata are not made of "springs and wheels," like those of Hobbes, but rather of logical rules, so their physical embodiment is largely irrelevant. But they

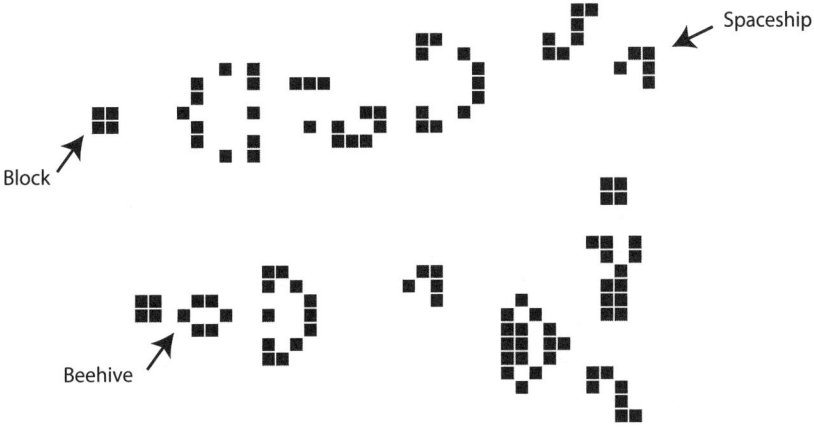

2.2 A snapshot of Conway's Game of Life.

can exhibit astonishingly complex sustained behavior, including self-replication and patterns resembling locomotion and evolution.

A famous example of a system of cellular automata is Conway's Game of Life, developed in 1970 by the British mathematician John Horton Conway. The game has a rectangular grid of cells that are either alive (shown as black squares) or dead (white squares). An initial state has some cells alive and some dead, as shown in figure 2.2. At each step of the game, the cells are updated according to the following rules:

1. Any live cell with fewer than two live neighbors dies.
2. Any live cell with two or three live neighbors lives on to the next step.
3. Any live cell with more than three live neighbors dies.
4. Any dead cell with exactly three live neighbors becomes a live cell.

Conway metaphorically associated these rules with life, where underpopulation, overpopulation, and reproduction could all change the state of a cell. Despite the simple rules, the game exhibits behavior that is not hard to interpret, at least metaphorically, as lifelike.

Cellular automata captured the imagination of many people and led to the development of many fascinating "digital organism" simulators much more complex than Conway's Game of Life. Most of these simulators share the principle of seeking lifelike complexity as an emergent property when many simple rules are applied repeatedly. According to the Danish theoretical biologist and philosopher Claus Emmeche, life can be calculated because life in itself realizes general forms of movement, forms of processing, that are computational in nature. If life is a machine, the machine itself can become living. The computer can be the path to life.[9]

Researchers started finding many complex processes in nature that seemingly could be explained this way. Inspired by the ability that simple rules have of generating complex patterns, some artificial life enthusiasts have even gone so far as to assert that all complex patterns in nature must have their origin in simple computational rules. One such enthusiast is Stephen Wolfram, whose monumental 2002 book, *A New Kind of Science*, concludes that "all is computation." Wolfram asserts that all natural processes can be constructed out of simple digital rules, even if we haven't yet figured out what the rules are. Complexities arise because of the chaos that such rules can induce. It is a big leap, however, to conclude that because there are computational patterns in nature, all patterns in nature are computational. I will examine later in this book many reasons why this may not be the case.

HELPLESS PROCREATION

I think we can agree that the ability to reproduce is necessary for any entity to be considered alive. Can an entity still be "alive" if it requires external help? Biological viruses require help from a host cell, computer viruses require help from a host program, and computer worms require help from an operating system. Even humans, however, require help from other living creatures in order to reproduce. For a woman to bear a child, she has to stay alive for nine months. Can she do this without the help of other living beings? What will she eat? And how will she digest what she eats without the help of her gut bacteria? Living for nine months is impossible without such help. So, in a sense, humans are not capable of fully autonomous reproduction either. Should we conclude from this that humans are not alive? Most creatures require *some* help from other living beings. Viruses and machines require more help than most others.

How does a computer program reproduce? There are many mechanisms, some quite familiar to all of us. When you go to an app store and install an app and then run it, have you just midwifed a new individual living digital being? When you start an app and it alerts you that a new version of the app is available, then, if you approve, the app downloads a mutant of itself and commits suicide. But you need not feel sorry for it. Suicide is also a routine part of biology. Biologists use the term *apoptosis* for cell suicide in multicellular organisms. Some fifty to seventy billion cells commit such suicide per day in the average human body. It is a normal part of a healthy organism and nobody feels sorry for these cells.

Analogies like apoptosis with software upgrades are admittedly quite a stretch, but they do suggest that understanding software as operating in a cooperative and competitive ecosystem may help us understand how technology evolves. Biological systems are the most complex dynamic systems on our planet today, but software systems are starting to get close in complexity to the simplest biological systems.

THE DURABLE AND THE DIGITAL

In an attempt to explain the complexity of biological systems, in 1944, Erwin Schrödinger, the Nobel Prize–winning Austrian theoretical physicist, wrote a landmark book entitled *What is Life?* Schrödinger, more than anyone else at the time, understood the workings of atoms, and he recognized that the traditional statistical tools of physics are inadequate to explain life. Such tools could explain how huge numbers of hydrogen atoms in the sun could generate heat and light, but not how a living cell could divide into two living cells.

In his book, published nearly ten years before Watson and Crick's paper describing the double helix structure of DNA, Schrödinger argued that life is rather a complex process that emerges from the interaction of much smaller numbers of atoms than those in the sun, each atom with its own function. This is not at all like the heat emerging from the sun as a consequence of statistical interactions of vast numbers of like elements, each contributing equally and identically. He recognized that the "chromosome fibre" played a central role and called it an "aperiodic crystal," arguing that the irregular structure of the molecules central to life encode the detailed functions of life. The mechanisms are not statistical in nature, but rather specific and operational.

In this regard, software is similar. The basic operation of a computer program is not a statistical outcome of billions of similar lines of code, but is rather a complex behavior emerging from the composition of a smaller number of lines of code, each with its own function. Just as a single line of code in a million-line computer program may prove essential to whether a machine lives or dies, a single misplaced atom in a DNA molecule may doom an organism. Life does, of course, rely on statistical properties, achieving robustness by having large numbers of redundant cells, for example. But at the level of biochemistry, "the working of an organism," Schrödinger asserts, "requires exact physical laws," not statistical ones.

This interplay of deterministic and exact operation with statistical emergent properties is echoed in technology. At the lowest level, a transistor is statistical, regulating the "sloshing of electrons,"[10] but above that level, a transistor is a precise and reliable digital switch enabling deterministic and exact digital operations on sequences of bits. At still higher levels, as with biology, statistical properties begin to dominate again. The Internet, for example, achieves robustness with redundant and self-adaptive routing of packets. Artificial neural nets, which have transformed technology by enabling image classification, speech recognition, and machine translation, to name a few examples, are inspired by the tangle of billions of neurons in the brain and rely on the aggregate effect of large numbers of simple operations.

Schrödinger also argued that the relative permanence of the molecules essential to life could not be explained by classical physics. Quantum phenomena, many of which Schrödinger himself had discovered, are essential. It is the quantum nature of atoms that makes it possible to *perfectly* copy a molecule, just as it is the digital nature of software that makes it possible to perfectly copy a program. And it is the quantum nature of atoms that makes molecules stable and durable, just as the digital nature of software makes programs durable.

AUTOPOIESIS

It is not just the molecules of life that are durable, but also their processes. Biological processes are self-sustaining. Physics and chemistry cooperate to create an entity, a living cell or an entire organism, that keeps the chaos and entropy of the world at bay, maintaining its own structure and, more importantly, its own activity. Google's servers, arguably, have similarly become a self-sustaining process. The processes even provide the mechanisms, serving up ads and billing clients, that entice their human symbionts, by richly remunerating them, to nurture, protect, and develop the servers and software.

In the 1970s, the Chilean biologists Humberto Maturana and Francisco Varela introduced the term *autopoiesis* to refer to a system capable of reproducing and maintaining itself. The term comes from the Greek *auto*, meaning "self," and *poiesis*, meaning creation or production, the root of the word "poetry." Here is Maturana's description of how the word came to him:

It was in these circumstances, while talking with a friend (José Bulnes) about an essay of his in which he analyzed Don Quixote's dilemma of whether to follow the path of arms (*praxis*, action) or the path of letters (*poiesis*, creation, production), and his eventual choice of *praxis* deferring any attempt at *poiesis*, I understood for the first time the power of the word "poiesis" and invented the word that we needed: *autopoiesis*. This was a word without a history, a word that could directly mean what takes place in the dynamics of the autonomy proper to living systems.[11]

An autopoietic entity is a network of processes that continuously regenerates and realizes itself.

The Blaster worm, like a biological virus, is a master of reproduction, with the help of a hijacked host, but also like a virus, it isn't much a process by itself. A Google server farm, on the other hand, is at least starting to resemble an autopoietic process, particularly when considered together with its human symbionts.

But we must be cautious. Analogies are a useful reasoning tool, providing "intuition pumps," to use the words of the philosopher Daniel Dennett, but Dennett warns us:

Analogies and metaphors. Mapping the features of one complex thing onto the features of another complex thing that you already (think you) understand is a famously powerful thinking tool, but it is so powerful that it often leads thinkers astray when their imaginations get captured by a treacherous analogy.[12]

Taking a cue from Dennett, while in this chapter I will focus on how digital technology resembles living beings, in chapters 7 and 8, I will consider important ways in which it does not. Those differences may ensure that AIs never actually resemble the humans of today. Humans of tomorrow may have merged with machines, in which case, resemblance may not be the right question. Do you resemble the trillions of bacteria in the micro-biome of your body? That question doesn't really make sense.

SPROUTING FROM TEENAGERS AND SPARKS

One problem with comparing computing systems to living ones is that that we don't really understand the mechanisms of biology as well as we understand the mechanisms of computing. For instance, while we have a pretty clear idea of how computing systems have come to be, our under-standing of how life first appeared is hopelessly incomplete.

But do we really understand as much as we think we do about the origin of computing systems? Computer viruses and worms, for example, are written by teenagers, so don't we need to understand teenagers to understand how these came about? That is probably about as hopeless as understanding how biological life came about.

The chemical reactions that constitute biological life today mostly involve proteins, which are composed of long chains of amino acids. In 1934, a graduate student at the University of Chicago, Stanley Miller, collaborating with his PhD thesis advisor Harold Urey, who had won the Nobel Prize for chemistry in 1934, conducted a series of famous experi-ments that showed how organic molecules essential to life could have first appeared.

Miller, who continued his work later at the University of California at San Diego, replicated in a sterile glass enclosure the conditions thought to be prevalent in the early Earth atmosphere (see figure 2.3). He showed that sparks in such an environment could trigger the synthesis of the same complex organic compounds that life depends on.

By itself, Miller's experiment did not demonstrate how these com-pounds could self-organize into the self-replicating systems that would lead to life. It only showed how the raw materials could have appeared. It was the work of lightning, not teenagers, according to Miller.[13]

More recently, some intriguing theories suggest that life may have come about more from self-organization of chemicals than from random reac-tions followed by natural selection, the mechanisms that Miller assumed dominated. The MIT physicist Jeremy England, for example, has devel-oped a model where random groups of molecules self-organize in order to

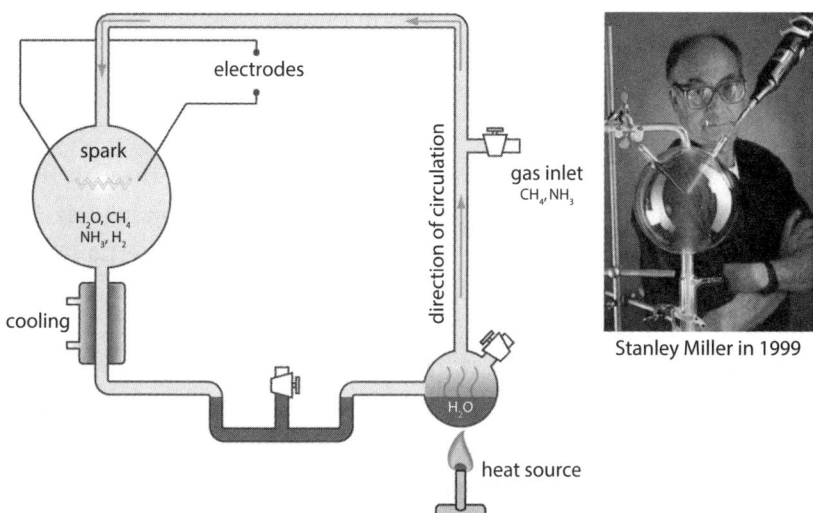

2.3 Schematic of the Miller-Urey experiment, which showed that amino acids essential to life could be synthesized by electrical sparks in gasses believed at the time to be common in the early earth atmosphere. By Carny at Hebrew Wikipedia. Transferred from the.wikipedia to Commons, CC BY 2.5.

more efficiently capture energy from the environment and dissipate it as heat.[14] His theory drastically reduces the importance of serendipity in the emergence of life. England has even shown how such self-organization can lead to self-replication, an essential property of living beings.[15] Such a theory may someday explain abiogenesis, the emergence of life from inanimate substances.

Another compelling theory is given by Stuart Kauffman, an American doctor who won the MacArthur "genius" award. Kauffman has proposed models where complex biological systems and organisms form as "attractors" in chaotic gene regulatory networks.[16] An attractor is a relatively stable operating mode into which a chaotic system can fall, and Kauffman has suggested that cell differentiation may come about as transitions between attractors. This could help explain why the same DNA can produce a heart cell and a hair follicle.

Whatever led organic molecules to self-organize and self-replicate, it is clear that those mechanisms were nothing like what is leading software to self-organize and self-replicate. The latter mechanisms have the heavy hand of humans, perhaps acting as the machine's gods. But the outcomes of these two mechanisms have more in common than one might expect. Humans, after all, are products of nature, so is not our hand in this also a product of nature?

Nature does produce beings other than humans that, through their actions, affect the course of evolution. For example, approximately 540 million years ago, an intense burst of evolution called the Cambrian explosion produced a very large number of multicellular species over a relatively short period of about twenty million years. In 2003, Andrew Parker postulated the "Light Switch" theory, in which the evolution of eyes initiated the arms race that led to the explosion.[17] Eyes accelerated evolution because they enabled predation. A predator facilitates the evolution of other species by killing many of them off, just as the sea kills boats. Is the hand of humans somehow different from the hand of predators? Are we at the start of a Googelian explosion?

REALLY LIVING

Max Tegmark is an MIT physics professor that, according to his Amazon author page, "is known as 'Mad Max' for his unorthodox ideas and passion for adventure." In his popular 2017 book *Life 3.0: Being Human in the Age of Artificial Intelligence*, Mad Max defines life broadly as a "process that can retain its complexity and replicate." He divides life on earth into three stages:

- Life 1.0: "evolves its hardware and software (biological stage)"
- Life 2.0: "evolves its hardware but designs much of its software (cultural stage)"
- Life 3.0: "designs its hardware and software (technological stage)"

He argues that we are entering the third stage because we are developing the technology now to manipulate our biological hardware, to extend our biological hardware with engineered devices, and to make software that designs the hardware on which it runs. By Tegmark's definition of life, digital machines are certainly living, but is his definition too broad?

Wikipedia, one of my favorite digital machines, in a wonderful page on "life," says, "the definition of life is controversial," and "it is a challenge for scientists and philosophers to define life."[18] It goes on that "life is a process, not a substance," and "any definition must be general enough to both encompass all known life and any unknown life that may be different from life on Earth." (Perhaps this requirement should be a bit stronger, to encompass any new life forms on Earth.) The page asserts that the "current definition" includes organisms that maintain homeostasis, are composed of cells, undergo metabolism, can grow, adapt to their environment, respond to stimuli, and reproduce (see figure 2.4).

Consider by this definition whether Wikipedia itself, or more specifically, the software system that serves pages and allows us to edit them, might be legitimately considered to be a living thing. Wikipedia has been continuously responding to stimulus from its (Internet) environment since 2001, when Jimmy Wales and Larry Sanger put the first version online. So at least it satisfies one of the seven requirements for living. It turns out that Wikipedia also entails processes analogous to the other six, at least insofar as these processes accomplish similar goals. The processes themselves, of course, have very different mechanisms, because mechanisms that work for organic chemicals do not work for electricity in silicon and vice versa. But please indulge me while I draw some parallels.

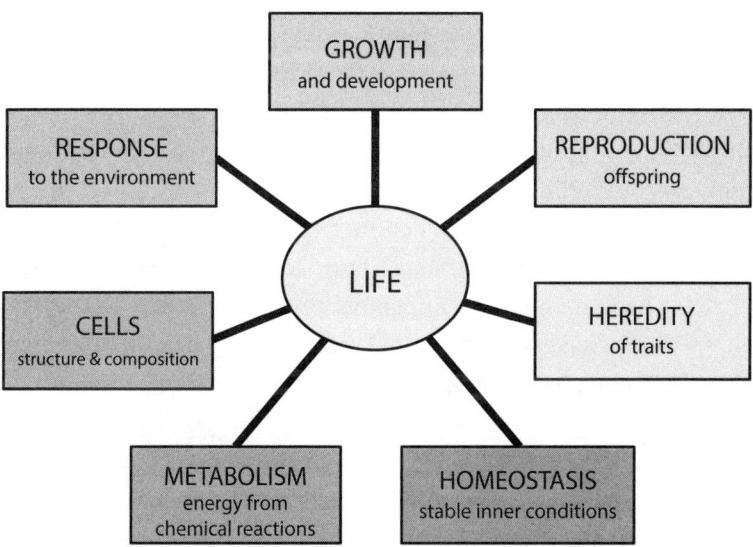

2.4 Properties of living beings. After Chris Packard, CC BY-SA 4.0.

FROM ORGIES TO EATING NATURAL GAS

I have already talked about reproduction. Except for computer viruses and worms, digital artifacts today mostly require human help to reproduce. But this is changing. Most computer programs are easy and cheap to copy and start a new execution, creating a new individual. Moreover, the copy can be exact, so heredity of traits is perfect. This form of reproduction is analogous to cell division, where two cells with the same DNA emerge from one. We also see more complex forms of reproduction. There is only one Wikipedia, for example, but many wikis that inherited essential features of Wikipedia.

What about sexual reproduction, where new DNA emerges from a random combination of two originators? Software engineers facilitate this form of software sex all the time. Very few software projects start with a blank slate and begin writing lines of code. Instead, engineers grab a piece of code from here and combine it with a piece of code from there. A typical program will "inherit" code from perhaps thousands of progenitors, a veritable orgy of sexual reproduction. The new program will inherit traits from each of the progenitors, but will become its own unique individual, ready to then replicate into thousands of perfect copies of itself.

For the most part, this sexual form of reproduction today requires help from humans, but that too is changing. Automated software tools can modify a program, for example to remove redundant operations, thereby creating a new mutant. And software that writes software has existed since the 1960s (in the form of compilers) with a steady evolution toward ever more abstract specifications forming the starting point for synthesis. There are even experiments today where machine learning algorithms drive software synthesis. So reproduction, heredity, and mutation of software has been trending for some time toward less human involvement. It is not hard to imagine shedding altogether the human role.

So it seems we have strong analogies to at least three of the seven requirements for life, responding to stimulus, reproduction, and heredity. What about the others?

Is Wikipedia composed of cells? Figure 2.5 shows racks of servers that the Wikimedia Foundation, a nonprofit organization headquartered in San Francisco, maintains to serve Wikipedia pages. Each server contains several processors, each of which could be considered a cell in a multicellular organism.

2.5 Wikimedia Foundation servers, which host Wikipedia pages. Photo by Victor Grigas/Wikimedia Foundation, CC BY-SA 3.0.

The Wikipedia processors have some of the same properties that cells in a biological organism have. For example, a processor can die without the overall organism dying. In fact, not one of the servers shown in the figure existed when Wikipedia was launched in 2001, yet Wikipedia has been in (approximately) continuous operation since then. The staff at the Wikimedia Foundation regularly replaces servers with newer models, and usually this does not require suspending the operation of the system. The removal of older and defective servers is perhaps an even better parallel to apoptosis than upgrading an app. If occasionally more major intervention is required, the system can be "put to sleep," so to speak, taken offline temporarily, not unlike a patient being put under anesthesia for major surgery. The patient temporarily stops responding to stimulus from the environment.

HOMEOSTASIS

Homeostasis is the maintenance of stable internal conditions. We mammals, for example, regulate our internal temperature using a variety of mechanisms including sweating. Other examples include maintaining blood glucose levels, blood oxygen, calcium levels, blood pressure, fluid balance, blood pH, and sodium concentration. Each of these mechanisms is accomplished with some form of negative feedback, where a sensor detects the level of the variable in question, and if the level is high, takes action to reduce it, and if the level is low, takes action to increase it. The pancreas, for example, secretes insulin to reduce blood sugar levels. If the blood sugar level is low, insulin secretion stops and alpha cells in the blood secrete glucagon, which causes glucose levels to rise.

Computers have fewer and simpler homeostatic mechanisms, but they most definitely have them. Every computer in the Wikipedia system includes a power supply that maintains stable voltage levels over a range of electric power inputs. The voltages coming out of a wall socket can vary quite a bit, spiking, surging, and sagging, but the power supply, using this sloppy electricity, feeds a much more stable DC voltage to the microprocessors. The microprocessors themselves include internally a number of voltage regulators that tolerate variability in the power supply levels and provide the internal circuitry with even more stable voltages.

Many computers also regulate the temperature of their "body." Modern data centers, like that shown in figure 2.5, have sophisticated air flow and air conditioning systems. Some modern microprocessors also internally regulate temperature, for example by slowing down if their internal temperature gets too high.

METABOLISM

Metabolism is more complicated. Metabolism is the collection of life-sustaining chemical reactions that occur within biological cells. Defined this way, no silicon-based digital technology has any form of metabolism. But if we look instead at what these chemical reactions accomplish rather than how they work, the parallel becomes stronger.

One of the functions of metabolism is to convert nutrients into energy. Computers get their energy from electricity, but where does the electricity come from? In the United States, it most likely comes from a power plant that "digests" natural gas in a chemical process. The "nutrients" for this process are organic molecules produced by biological organisms, just like the food we eat, although computers' "food" is much older than anything we would eat. So if we are willing to consider the power plant to be their digestive system, then the analogy becomes a bit less farfetched.

Electricity is a more directly usable energy source than, say, sugars, which are a primary source for living cells. In figure 2.5, a backplane delivers electricity to each cell, whereas in a mammal, for example, a circulatory system delivers nutrients and oxygen to the cells, and the cells in turn metabolize those nutrients to convert them to energy. Wikipedia's servers would be closer to the biological design if the backplane were to deliver natural gas to each server, and then the server, using a small power plant, were to generate electricity locally.

Machines occasionally "starve," as batteries run out or hurricanes cause power failures. A system like Wikipedia, however, is less likely to starve because it is geographically distributed, with servers scattered across the globe. If one server farm goes down, the others can pick up the load, albeit with increased delay. This mechanism, the dynamic rerouting of queries to those servers that remain working, can be viewed as a form of self-repair, a feature of many biological living beings, though not one included as a requirement in the Wikipedia page for "life."

GROWING

As of this writing, Wikipedia is eighteen years old. By any measure, it has grown enormously in that time. Its physical body in 2001 consisted of a single server, whereas by 2019, the Wikimedia Foundation was maintaining five server farms like that shown in figure 2.5, three in the United States, one in The Netherlands, and one in Singapore. When you visit Wikipedia, your browser will be directed to the nearest of these five data centers.

Wikipedia's growth has been driven by humans. It has not been autonomous. But the growth of many biological beings, including humans, also depends on other living beings. If machines are symbiotically intertwined with humans, then it would be natural for their growth to depend on humans.

This analogy with biological growth is admittedly a bit tortured, but consider that technology today is developing rapidly. Manufacturing plants that make the computer chips, printed circuit boards, power supplies, and enclosures that form the "cells" of Wikipedia are increasingly computer controlled. Bostrom and Tegmark both postulate that eventually the computers will take control of the manufacturing processes that make their own components. If this happens, the analogy will become much less tortured.

BRAINS, MINDS, AND THE SKY

So it seems that we have reasonable parallels for all seven requirements for life, plus at least one more, self-repair. There are additional properties found only in more advanced biological life forms, most notably nervous systems and cognition. Many digital artifacts lack anything analogous to these higher-level functions, but this, too, is changing. Wikipedia most certainly incorporates something similar to a nervous system. The wires in figure 2.5 are Ethernet cables linking the servers to each other and to the Internet. The five Wikipedia sites are also linked to each other through the Internet. Is this not analogous to a nervous system?

Having a nervous system enables communication between components. In biology, a sufficiently complex nervous system also enables cognition and consciousness. But is it only a human-like machine that can have these properties? The Australian philosopher of science Peter Godfrey-Smith, in his wonderful 2016 book *Other Minds: The Octopus, the Sea, and the Deep Origins of Consciousness*, questions our deeply anthropocentric and unitarian view of consciousness. The most fascinating part of the book is his study of octopuses, which have evolved brains quite independently of humans. Our latest common ancestor had nothing of the sort. The brain of an octopus is spread throughout its body, so its architecture is very different from ours. Yet an octopus exhibits distinct signs of intelligence, self-awareness, and consciousness. Almost certainly, its experience of these phenomena is very different from our own experience, but it shows that machinery quite different from our own can manifest cognitive functions that seem to resemble ours. Does Wikipedia have anything resembling cognition or consciousness? Will any future digital technology develop such features? This question would be much easier if we had a clear idea of what cognition is, but such clarity is likely to remain elusive.

I am now going to put a stake in the ground and take a solidly materialist stance. Materialists believe that mind and consciousness are byproducts of biochemistry and the brain's physical interaction with its body and its environment through sensors and actuators. If instead we humans have an immaterial soul or any other nonphysical origin of mind or consciousness, what Daniel Dennett calls "wonder tissue,"[19] then I do not know how to even address the question of whether silicon-based machinery and software could have such a soul. Without a materialist stance, you would need to consult your priest, rabbi, shaman, guru, or other spiritual leader on this question. I am not a spiritual leader and therefore not qualified to address this question from that perspective. So regardless of your own beliefs, it only makes sense to interpret what comes next through the lens of a materialist.

A remarkably prescient materialist was the American poet Emily Dickinson (1830–1886), whose poems make frequent reference to the brain

2.6 Daguerreotype of Emily Dickinson, c. early 1847, taken by an unknown photographer. This photo is presently located in Amherst College Archives and Special Collections.

in places where other poets would have put the soul. One such poem is this one:

The Brain—is wider than the Sky—
For—put them side by side—
The one the other will contain
With ease—and you—beside—

The Brain is deeper than the sea—
For—hold them—Blue to Blue—
The one the other will absorb—
As Sponges—Buckets—do—

The Brain is just the weight of God—
For—Heft them—Pound for Pound—
And they will differ—if they do—
As Syllable from Sound—[20]

The Harvard cognitive psychologist Steven Pinker, who appeared at the end of the previous chapter, comments on this poem:

The first two verses of Emily Dickinson's "The Brain Is Wider Than the Sky" express the grandeur in the view of the mind as consisting in the activity of the brain. Here and in her other poems, Dickinson refers to "the brain," not "the soul" or even "the mind," as if to remind her readers that the seat of our thought and experience is a hunk of matter. Yes, science is, in a sense, "reducing" us to the physiological processes of a not-very-attractive three-pound organ. But what an organ! In its staggering complexity, its explosive combinatorial computation, and its limitless ability to imagine real and hypothetical worlds, the brain, truly, is wider than the sky. The poem itself proves it. Simply to understand the comparison in each verse, the brain of the reader must contain the sky and absorb the sea and visualize each one at the same scale as the brain itself.[21]

The brain links a physical phenomenon, the sky, with a concept, in such a way that conjuring the concept of "sky," an act that is nothing more or less than a pattern of neurons firing, is linked to the pattern of firings that constitute the visual perception of a sky. Linking patterns is the essence of Wikipedia, so to the extent that it is the links themselves that constitute a cognitive grasp of a concept, then it certainly seems plausible that Wikipedia has a powerful grasp on many intellectual concepts that matter to humans, such as the sky.

As of this writing, however, Wikipedia is largely limited to patterns of language, words and characters, so its concept of a "sky" cannot possibly

2.7   Sky. By Jessie Eastland [CC BY-SA 4.0], from Wikimedia Commons.

match ours. Although Wikipedia includes many images (see, for example, the sky image in figure 2.7, taken from the Wikipedia page on "sky"), it is not yet very good at linking and searching images. This is changing, however. Automated image understanding has seen spectacular success in recent years. And several websites support searching for images that are similar to some given image. It is not farfetched that in the near future, clicking on the clouds in figure 2.7 will take us to the Wikipedia page on "cloud."

The third verse of Dickinson's poem invokes spirituality, but in a rather odd way. Pinker comments on this verse:

The enigmatic final verse, with its startling image of God and the brain being hefted like cabbages, has puzzled readers since the poem was published. Some read it as creationism (God made the brain), others as atheism (the brain

thought up God). The simile with phonology—sound is a seamless continuum, a syllable is a demarcated unit of it—suggests a kind of pantheism: God
is everywhere and nowhere, and every brain incarnates a finite measure of
divinity. The loophole "if they do" suggests mysticism—the brain and God
may somehow be the same thing—and, of course, agnosticism. The ambiguity
is surely intentional, and I doubt that anyone could defend a single interpretation as the correct one.[22]

CONNECTIONS

The idea that connections or links are everything in cognition has sparked
a new field of neuroscience called connectomics, which is explained by
the Harvard neuroscientist Jeff Lichtman (see figure 2.8) as follows:

The brain's structure is more complicated than that of any other known biological tissue. As a result, much of the nervous system's fine details, such as the
vast neuronal circuits that connect nerve cells together at synapses are largely
unexplored. My colleagues and I have developed automated methods to both
generate and analyze digital data sets that reveal all the neuronal wiring and
many subcellular details of brain tissue. We use a novel means of cutting brains
into very thin slices, and a new electron microscope that acquires images of the
brain at unprecedented speed and resolution, so that in a volume of brain, every
synaptic connection between nerve cells is visible. These acquired data sets are
very large: a cubic millimeter of brain requires acquiring more than 2 million
gigabytes of image data. The brain reconstructions coming out of this work
reveal networks that are even more complicated than we imagined. In our view,
this new approach (which we have dubbed "connectomics") shows promise to
be sure; nevertheless, many challenges remain. Most serious of these may be a
fundamental limit to what our human brains can understand.[23]

The key idea behind connectomics is that a map of the connections
between neurons in the brain will somehow lend insight into how the
brain works. In a way, the spectacular work of Lichtman's lab shows that
such insight may be hard to come by. The staggering complexity of the
map makes it unlikely that its structure will explain anything, at least
with the current state of technology for acquiring and analyzing the
data.[24]

Before I heard of Lichtman's work, I had a profound misconception
of the structure of a brain, a misconception formed from having seen
many images of neurons that were made using Golgi's method, named

2.8 Jeff Lichtman of Harvard presenting a computer-generated, three-dimensional reconstruction from images of tiny slices of a small section of a brain. Image from iBiology.org. Reproduced with permission.

after Camillo Golgi, an Italian physician who published the first picture made with the technique in 1873. Golgi's method was used by the Spanish neuroanatomist Santiago Ramón y Cajal (1852–1934) to expose, for the first time, the structure of neurons. The images created by Ramón y Cajal and many others since show spindly dendrites with vast empty space between them (see figure 2.9). These images are misleading because Golgi's method stains a tiny fraction of the neurons, perhaps 0.1 percent, leaving the others unseen. Lichtman's team showed that all that empty space is packed with thousands of other tangled neurons.

The number of links in Wikipedia's millions of articles pales in comparison to the number of links in a single human brain. This is true whether we consider the physical links, the Ethernet cables, or the hyperlinks, which link concepts. There are many orders of magnitude fewer connections, by any measure, in all of Wikipedia than in a single human brain. But is the difference just a matter of scale? If Wikipedia grows to have a comparable number of connections, will its operations start to more closely resemble those of a brain? In subsequent chapters, I examine this question.

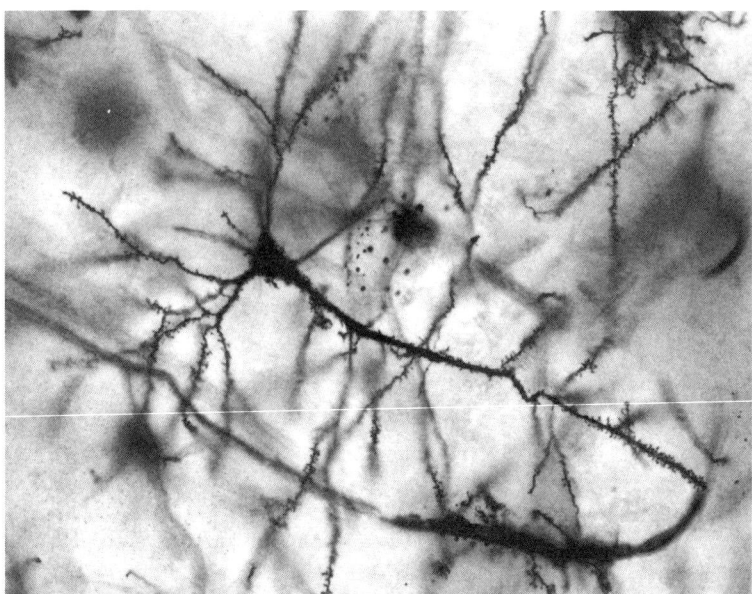

2.9 Golgi-stained neuron in a human hippocampus. By MethoxyRoxy, CC BY-SA 2.5, via Wikimedia Commons.

LEARNING, PAIN, AND PLEASURE

Many biological life forms learn. A biological organism will, if possible, avoid taking an action that has previously caused it pain, and will strive to repeat an action that has previously caused it pleasure. The neurobiological phenomena of pleasure and pain evolved precisely because they reinforce connections in the brain, producing the memory needed to affect future actions.

Machines can learn too. Machine learning, a branch of statistics and computer science that is viewed by many people as a subfield of AI, has existed at least since the 1950s. Most of us rely on algorithms developed in this field to identify spam email and sequester it in a folder labeled "spam." Optical character recognition, widely used today to process checks and to convert legacy documents for online use, is another success case for machine learning. Image classification and machine translation are two examples that have seen spectacular improvements in recent years.

What really is machine learning? The Carnegie Mellon computer scientist Tom Mitchell offered a widely used definition in a classic book on the subject:

A computer program is said to learn from experience E with respect to some class of tasks T and performance measure P if its performance at tasks in T, as measured by P, improves with experience E.[25]

By this definition, does Wikipedia learn? To answer this question, we need to define experience E, tasks T, and a performance measure P. Anything "experienced" by Wikipedia must take the form of stimulus from outside the system. The outside stimulus to Wikipedia has two forms, one is page views and the other is page edits. Both come in through the Internet.

Let's first consider the simpler of these, page views. To determine whether Wikipedia learns anything when a user follows a link to a page and views it, we still have to define tasks T and a performance measure P. To keep it simple, let P be the average response time, the time it takes between when the user clicks on a link and the page has been served to the user's computer. This is an easily measured performance measure. The tasks T, then, are just further stimulus of the form of page views. According to a Wikimedia blog published on April 24, 2018,[26] by these definitions, Wikipedia does learn from page views. Recall that I mentioned that Wikimedia maintains five data centers around the globe to serve pages. The pages are stored in a centralized database that keeps the master copy, but each data center maintains a cache with copies of the most frequently accessed pages. When you click on a link, your computer will talk to whichever data center is physically closest, and if that data center happens to have the page you are asking for in its cache, it will quickly respond with the contents of the page. Otherwise, it has to query the central database to obtain the page, and the response time will be longer. But in this case, it will save the page in its cache, so when your neighbor goes to access the same page, the response is quicker. Each page view teaches the data center a little bit about which pages are accessed most often, and hence the system learns.

The purpose of Wikipedia is to gather the collective wisdom of humanity. This seems like a far more interesting form of "learning" than what I have just described. This is true, but it's more challenging to come up with a performance measure P for this form of learning. We could try letting P measure the fraction of humanity's knowledge that has been captured on Wikipedia pages, but I wouldn't have any idea how to measure that. We could try letting P measure the accuracy of Wikipedia pages, but I suspect that for many pages, accuracy will be determined by opinion rather than by objective fact. We could measure how happy Wikipedia users are with Wikipedia, but it would be costly to conduct a proper survey. Perhaps we could measure this happiness indirectly, for example by the amounts of donations to the Wikimedia Foundation. I personally have been increasing my donations every year because I have been finding the resource more valuable every year. We could also try to measure how much writers of books like this one use Wikipedia. I personally am using it extensively for this book, though I try to be careful to corroborate the information I get. After all, anybody can edit a Wikipedia page. But this is anecdotal evidence, not a measure of performance. We could measure the happiness of Wikipedia users by measuring the number of unique page views per unit time. In this case, the experience E is page edits, the task T is to serve pages, and the performance measure P is the number of page views per unit time. I have no idea how to prove it, but it seems obvious that P has increased with more E. If there had been far fewer page edits, I doubt I would visit Wikipedia as often. Intuitively, it seems that Wikipedia is learning about the totality and connectedness of human concepts, but using Mitchell's definition of machine learning, I cannot rigorously defend that intuition.

Nevertheless, I have no doubt that Wikipedia is learning in a distinctly cognitive sense. By May 2018, there were well over five million Wikipedia articles in English, and about six hundred new articles were being added every day, each with many links to prior pages and to outside resources. Wikipedia is learning from its human progenitors. We teach it how ideas are interconnected. In return, it rewards us by serving as a cognitive prosthesis.

But perhaps this is all wrong. The mechanisms that make digital machines work are very different from those that make us work. I will return to this counterargument in chapters 7 and 8. But first, let us dive more deeply into the question of cognitive prostheses. Even digital technology that falls far short of anything we would call intelligent can affect the intelligence of humans, both positively and negatively. This is the topic of the next chapter.

3

ARE COMPUTERS USELESS?

FLYNN'S IQ

I had an argument once with my daughter's tenth-grade teachers. My daughter was confused by the Bohr model of the atom, and I suggested that the Wikipedia page on the subject looked to me like an excellent source. They were not enthusiastic. "We do not encourage that." The textbook, with its slick illustrations, should be all she needs. "But," I said, "sometimes it helps to hear it said another way."

"We do not encourage that."

I think the real issue wasn't having it said another way; it was having it said by Wikipedia. It's an anarchy of ideas, isn't it? Anyone can edit those pages.

And yet, much to the astonishment of many, that anarchy turns out to be a pretty good repository of human knowledge. How can that be, in a world where teenagers create viruses for fun? Despite the potential for chaos, I'm convinced that Wikipedia makes us smarter. Somehow, a culture has emerged that values improving on each other's work. Many Wikipedia pages are not so good, but some are spectacular. I have, upon occasion, found a better description of a topic I know well on Wikipedia than anywhere else. Every time that has happened to me, I've been awed. Wikipedia has become my first go-to source for just about anything mathematical, for example. Perhaps these pages have the advantage of being more checkable than, say, a page on a controversial political topic. When errors creep into a page on a mathematical topic, they are more evidently errors.

I am old enough to remember the days when I would need to make a trip to the library to find a book that would explain a mathematical concept that I needed to know. Now, I can go to Wikipedia, collect a few ideas, then go to Google Scholar and sort through the relevant papers. The cost in time and effort to find information used to be much greater, and as a consequence, we did less of it. I'm convinced that Google Scholar and Wikipedia make me smarter, but what does that mean?

Intelligence is hard to measure, but one fairly well-established method is IQ tests. It turns out that during much of the twentieth century, performance on IQ tests steadily increased at a rate of about three IQ points per decade. This phenomenon has been called the Flynn effect, named after James Flynn (born 1934), a New Zealand professor (emeritus) of political studies who wrote extensively about the effect. IQ tests are normalized, so the average score across a population is always set to 100. But the tests are continually redesigned, and for a period of time, when new test takers took older tests, their average scores were significantly above 100. In a 2013 TED talk, Flynn says that if you score people of a century ago using modern IQ tests, they would have an average IQ of 70, a level that today is considered a serious intellectual disability. If you score people today using the standards of a century ago, they would have an average IQ of 130, a level that today is considered gifted. Flynn argues that this phenomenon is a result of the much more complex world that we live in. The increased complexity of the world we live in is due, at least in part, to technology and the more complex social structures that have been enabled by technology. Is it possible that technologies like Wikipedia actually make us measurably smarter?

Unfortunately, in recent years, coincident with the rise of Wikipedia, the rise in IQ scores has slowed and even reversed in several Western countries. Interestingly, experts cite technology as contributing both to the Flynn effect and to its reversal.[1] For the twentieth-century rise, perhaps a more plausible explanation is better nutrition, ubiquitous education, better health, and a cleaner environment. Cultural complexity appears to continue to rise while IQ scores fall, so it seems an unlikely explanation for the Flynn effect.

But IQ tests, by design, are attempting to measure the capabilities of an individual human brain. The test takers are allowed to use old-fashioned cognitive prostheses, paper and pencil, when taking the test, but they are prohibited from using modern cognitive prostheses like Google search. And the tests do not even attempt to measure collective intelligence. Better-networked humans can solve more problems effectively than isolated humans. Perhaps cultural complexity is part of a rise in collective intelligence, not individual intelligence.

IQ RISING, BRAINS SHRINKING

Kevin Laland, professor of behavioral and evolutionary biology at the University of St. Andrews in Scotland, in his book *Darwin's Unfinished Symphony*, states that, "human minds are not just built for culture; they are built by culture," and "culture is not just a product, but also a codirector, of human evolution."[2] Laland, however, is talking about evolutionary time scales, many thousands of years, not the blink of an eye covered by the Flynn effect. Laland chronicles the correlation between brain sizes of animals and the complexity of their social structure and points out that "increases in hominin brain size coincide with advances in technology."[3] Indeed, brain sizes of humans and our ancestors have increased dramatically over the last two million years or so, but curiously, average human brain size has *decreased* by about ten percent in the last ten thousand years, a period coinciding with the most rapid technological advancement. Christopher Stringer, a paleo-anthropologist and research leader on human origins at the Natural History Museum in London, suggests that technology may make larger brain sizes unnecessary:

The fact that we increasingly store and process information externally—in books, computers and online—means that many of us can probably get by with smaller brains.

He goes on:

The way we live may have affected brain size. For instance, domesticated animals have smaller brains than their wild counterparts probably because they do not require the extra brainpower that could help them evade predators or hunt for food. Similarly, humans have become more domesticated. But as long as we keep our brains fit for our particular lifestyles, there should be no reason to fear for the collective intelligence of our species.[4]

Wikipedia and other digital technologies, apparently, have become extensions of our brains and perhaps, according to Stringer, have physically replaced previously biological mass in the brain, all the while making us smarter, at least collectively. Technology certainly enables a more complex social structure, and this complexity drives the brain to adapt to handle it. Technology augments our brains with additional, usable hardware and processes.

Under this reasoning, we are already cyborgs. Just as we delegate parts of digestion to our gut biome, we delegate parts of our thinking to technology. Like our gut biome, digital technologies are dynamic processes, not passive inert physical artifacts. That they are dynamic and have become a part of us may be reason enough to regard them as living.

MASSAGING THE MESSAGE

In the 1960s, long before the Internet and the World Wide Web, the Canadian English professor Marshall McLuhan began a controversial exploration of the idea that technology for communication between humans, such as printing, television, and radio, shape not just culture, but also our cognitive selves.[5] If media technologies are, as McLuhan asserts, extension of our selves, then what would he make of artificial intelligences?

McLuhan anticipated many aspects of our digital culture. He predicted that electronic media would push our society from individualism to a collective identity, what he called a "global village." If he were alive today, I'm certain McLuhan would have a lot to say about the collective collaborative work of Wikipedia. McLuhan, anticipating the effect that the web would have on our way of reading, absorbing information, and even thinking, popularized the term "surfing" for rapid random movement through a heterogeneous body of documents. When he famously said, "The medium is the message," he claimed that the structure of the medium itself, rather than just its content, shapes our thought. Television, its broadcast format, its interruptions for advertisement, its color and style, became part of what it meant to be a twentieth-century person.

The structure of today's media, however, is vastly richer than what McLuhan knew. I suspect that Facebook, Twitter, and Google would have blown his mind because those media actively shape us as if they have their own agency. As the 2016 US presidential election taught us, social media, using personalization algorithms that watch us and learn our predilections, create for each individual an echo chamber, where our ideas are reinforced and opposing views are never seen. For shaping our cognitive selves, I believe, these algorithms go much further than anything McLuhan predicted. These media shape thought in ways that even the algorithm designers at Facebook could not have predicted. While technology may be making us collectively smarter, it could actually be simultaneously making us individually dumber.

ISLANDS OF DISJOINT TRUTHS

Pierre Teilhard de Chardin, a French philosopher and Jesuit priest, developed the idea of the "noosphere" in the 1920s to describe the third phase of development of the Earth after the geosphere (inanimate matter) and the biosphere (biological life). The personalization algorithms in social media have fragmented Teilhard de Chardin's noosphere into islands of disjoint truths. Just as the biosphere partitions organisms into species that cannot interbreed, social media partitions knowledge into noospecies that do not even share basic truths. Incompatible world views fester.

Teilhard de Chardin was fundamentally a spiritualist, and his noosphere was not a property of either the geosphere or the biosphere, but existed separately, distinct from any material reality. He had a complex relationship with the Catholic Church, which included many of his writings in the *Index Librorum Prohibitorum* (the official list of prohibited books). He participated in the expedition that discovered Peking man, a group of fossils of human ancestors from some 750,000 years ago, and he lectured about evolution, directly contradicting established dogma of his time. But he was posthumously praised by Pope Benedict XVI and cited by Pope Francis in the 2015 encyclical for his theological writings.

If the noosphere is not a property of the physical and biological world, then would Teilhard de Chardin have allowed digital machines into his spiritual world? It seems unlikely, even heretical, but humans do not seem to require consistency in their islands of truth. Evangelical Christians in the United States, for example, overwhelmingly supported Donald Trump in the 2016 election despite his many incompatibilities with the teachings of Jesus and the traditions of Christianity.

Today, we are making machines that arguably possess some form of "knowledge" and can arguably form some kind of "belief." I could take a humanist perspective, and argue that when pundits today assert that we are in a "post-fact" world, then we need to re-examine what we mean by "knowledge" and "fact." And when entire populations overtly act against what they say they believe in, we have to re-examine what we mean by belief. But I will instead take a technologist's perspective. If we can make machines that know things and have beliefs, then maybe we can learn something about what knowledge and beliefs are. We can ask, for example, whether the machine form of knowledge will be any more consistent than that of humans.

It may well be that machines are better at knowledge and belief than we are. Barrat, Bostrom, and Tegmark seem sure that the machines will outstrip human capabilities in every dimension, including knowledge and belief. However, I think the question is more nuanced than who will win the race, man or machine. Cognitive prostheses like Wikipedia and Google are forces pointing toward a symbiosis with technology, not a competition. There will be stresses, even disasters, but they will be more like diseases than attacks. Our society and culture, intertwined with the machines, will suffer illnesses, and I hope we will overcome them. But these problems will be pathologies, not a War of the Worlds.

Pathologies, like wars, can kill, so they must be taken seriously. The computer scientist and artificial intelligence (AI) pioneer Stuart Russell points out that the situation is even worse because the curated information that the AIs feed to you *changes you*.[6] The algorithms are designed to maximize click-through, the probability that you will click on presented advertisements. The algorithms not only adjust what they present to you based on their prediction of what you are likely to click on, but also feed you information that makes you more likely to click on whatever they predict you will click on. This is a positive feedback loop (see chapter 5) that tends to drive people toward polarization. Political extremists are more predictable than moderates. More generally, people with narrower world views are more predictable, so more sharply delineated islands of disjoint truths make the algorithms more effective.

There can be no doubt that technology is shaping our cognitive selves, but neither McLuhan nor Teilhard de Chardin could have anticipated that it would be doing so through cognitive functions of its own. The machines observe, learn, and reflect human thought, and then synthesize structures of knowledge and subcultures. A Google search, when it displays the top ten hits for a phrase, gives a new meaning to that phrase, linking it to concepts that no human would have done. We simply don't have the capacity to recall so many things. The machines are already deeply intertwined with cognition, and yet everyone knows that AI is in its infancy. Where, I have to ask, are we headed?

I can summarize the grim situation as follows. If we can call the organized data and trained neural networks in the machine "knowledge," then the machines know vastly more than any individual human could possibly ever know. The Israeli historian Yuval Noah Harari, in *Homo Deus*, defines "dataism" as the religion of the day and says these AIs are coming to know us better than we know ourselves.[7] They use that knowledge to curate individual information flows. As Russell points out, their algorithms tend to craft those information flows to narrow rather than widen the individual's worldview, reinforcing their predilections and making them more predictable. The result is a noosphere, to use Teilhard de Chardin's word, composed of festering islands of disjoint truths. Humanity divides into cults of radical religious fundamentalism, white supremacism, far left, reactionaries, Trumpism, conspiratorial anarchists, and many other worldviews that lie outside my own island of truths and therefore will forever remain incomprehensible to me. Can democracy survive this?

The serial entrepreneur Elon Musk stated in 2017 that AI represents an "existential threat to humanity." But I don't think we are on the precipice of annihilation by a hostile silicon-based civilization, as depicted in so many science-fiction movies. The threat is coming from within, as we humans change with the technology. Many of us, as individuals, are facing an existential crisis, but it is not annihilation we face (at least not by the AIs), it is metamorphosis. We may well wake up tomorrow as a cognitive cockroach.

COGNITIVE COCKROACHES

In his 2018 book, *21 Lessons for the 21st Century*, Harari predicts that while the human struggle against exploitation dominated the twentieth century, in the twenty-first century, it will be a struggle against irrelevance. He argues that people will need to worry more about not being needed than about being used. It is not hard to envision many consequent societal pathologies that will have to be managed. Once the algorithms determine the outcome of democratic elections, for example, then the actual act of voting becomes a farce.

In Franz Kafka's *The Metamorphosis*, first published in 1915, a traveling salesman named Gregor Samsa wakes up one morning having transformed overnight into a giant insect. He lies in bed, staring at the ceiling and flailing his multiple legs, contemplating how he has overslept and missed the train for work. Gregor's office manager shows up and informs Gregor through the closed door that missing work will have dire consequences, particularly in view of his recent poor performance. When Gregor finally manages to open the door, the office manager, horrified by the giant insect, flees, and Gregor becomes officially jobless. He becomes a burden on his family, no longer able to provide for them, falling into a frightful form of Harari's irrelevance. Gregor languishes and eventually dies.

Gregor's career, traveling salesman, no longer exists. Some interpretations of Kafka's story see Gregor's grotesque physical metamorphosis as a graphic representation of his transformation from a stable family provider to an indolent parasite, a burden on the family. This transformation happened through no fault of his own. Many people today work in careers that are threatened by technology and could wake up tomorrow finding themselves suddenly a burden on their families.

The American futurist Martin Ford, in his 2015 *New York Times* bestseller, *Rise of the Robots*, presents a rather pessimistic view of the future of

3.1 Cover of the 1916 edition of Franz Kafka's *The Metamorphosis*.

work. He argues that the current phase of technological change is qualitatively different from the increased automation of previous generations in that it is now knowledge workers who will be displaced. Many knowledge workers may indeed wake up as cognitive cockroaches, unable to provide for their families. In Kafka's story, Gregor changed while his world remained the same, while in Ford's story, Gregor would be unchanged while his world morphed. Either way, Gregor becomes unable to function in his new world.

Ford is unquestionably correct that we are in for big changes and that the lives of many individuals will be disrupted, some in tragic ways like Gregor's. But it's also possible that life around those individuals will go on, and maybe thrive, even if transformed. Kafka's story ends, after Gregor's death, with Gregor's parents noticing that his sister Grete has grown up into an attractive young lady and that they should begin to think of finding her a husband. Our children, growing up with the new technology, will define the new normal, whatever that ends up being. It may be dystopian, but humans have historically proved to be robust and adaptable.

There is room for cautious optimism, particularly if we understand what is going on. Instead of being displaced by technology, the more adaptive humans among us may be enhanced by it. This does not diminish the tragedy of those who fail to adapt, but makes room for optimism about humanity as a whole. Perhaps a cognitively enhanced humanity can even learn to avoid being stranded on islands of disjoint truths. And perhaps we can learn more humane ways to deal with individuals who fail to adapt than what befell Gregor.

CAUTIOUS OPTIMISM

In his book, *Homo Deus: A Brief History of Tomorrow*, Harari pulls no punches. He states that "humans are in danger of losing their economic value because intelligence is decoupling from consciousness."[8] But he also points out that recent technological advances appear to have had a profoundly salutary effect on humans:

For the first time in history, more people die today from eating too much than from eating too little; more people die from old age than from infectious diseases; and more people commit suicide than are killed by soldiers, terrorists, and criminals combined.[9]

Harari points out that the ancient Sumerians, some five thousand years ago, invented the concept of money and a writing system, one focused on numbers and bureaucracy, and "broke the data-processing limitations of the human brain." The consequences were profound:

Writing and money made it possible to start collecting taxes from hundreds of thousands of people, to organise complex bureaucracies and to establish vast kingdoms.[10]

Armies and corporations emerged with intelligence but no consciousness. Enabled by a technology for writing, which makes possible recording and manipulating numbers, laws, and contracts, these social constructions

3.2 Fragment of Sumerian writing on a clay cone from about 2350 BC, currently located in the Louvre Museum. The inscription documents some accomplishments of a Sumerian prince.

have enabled the expansion of humanity to seven and half billion individuals on this tiny planet. The magic ingredient, according to Harari, is communication between people in ways that our biology alone does not support:

Over…20,000 years humankind moved from hunting mammoth with stone-tipped spears to exploring the solar system with spaceships not thanks to the evolution of more dexterous hands or bigger brains (our brains today seem actually to be smaller). Instead, the crucial factor in our conquest of the world was our ability to connect many humans to one another.[11]

I recall being surprised, sometime in the mid-1980s, when computers switched from being number crunching devices to being communication media. Pablo Picasso is said to have said about computers, "But they are useless. They can only give you answers." They were developed, after all, to simulate nuclear chain reactions. But the computers Picasso knew were not the computers of today. Computers today connect people and their ideas in ways that no amount of Sumerian writing could.

Interconnected computers emerged in the 1980s, less than forty years ago. This is a blink of an eye by historical terms, and yet the profound and rapid change they have brought about is necessarily alarming. Are computers going to evolve into human-like thinkers, into all-powerful superintelligences, into tools that augment human thinkers, or into none of the above? A first mistake we often make when discussing such questions is to assume that humans will not evolve along with the technology. We imagine superintelligences interacting with our culture of today, and it becomes easy to imagine a dystopia. But if McLuhan is right about anything at all, then humans most certainly will change with the technology, as we always have. I know that I have changed. I never could have written this chapter, featuring McLuhan, Picasso, Teilhard de Chardin, Trump, Musk, Jesus, Harari, Ford, and Kafka without Google, Wikipedia, searchable e-books, and my trusty MacBook Pro with its Spotlight feature that can find my notes. These technologies are extensions of my brain, and as is probably evident to you from what you are reading, they have created for me a truly bizarre echo chamber.

Wikipedia is not an AI, but it enhances the intelligence of humans. Since the invention of writing by the Sumerians, technology has done that without requiring any intelligence of its own. A clay tablet has no intelligence. Neither is intelligence of its own required for us to consider something to be alive. We do not demand intelligence of our gut biome. Nevertheless, while intelligence is not required to *be* alive, it can certainly help for *staying* alive. For this purpose, even a small amount of intelligence can go a long way. A small amount of intelligence, like that found in a worm, cannot be measured by IQ tests. It can't even be defined simply. In the next chapter, I will argue that the most elemental forms of intelligence involve feedback and do not require consciousness, a theme that will develop over the next few chapters.

4

SAY WHAT YOU MEAN

DID I SAY THAT?

I do quite a bit of public speaking. I have tried writing a script and memorizing it, but that rarely works. I think the only time it worked for me was when the US National Science Foundation asked me to give a ninety-second talk about my research. It is surprisingly difficult to say so little when you want to convey so much, and I couldn't figure out how to stay within the ninety-second limit without a word-for-word script.[1] On a few other occasions, I tried writing a script for a longer talk and reading from it. This was a disaster, yielding a stilted, soporific speech. So most of the time, I do not plan what to say. I have a general idea, of course, and I use PowerPoint slides to prompt me and provide visual stimulus to the audience, but what words will come out of mouth, I cannot predict. I have to trust my brain to synthesize the right words on the fly. Most of the time, it does a reasonable job.

Anyone listening to one of my talks, rightly, holds me accountable for my words. But did I consciously choose those words? If I am reading from a script, then I would have to say, yes, I consciously chose those words. But with impromptu speaking, I become consciously aware of the words I have strung together only after I have said them, or, perhaps, as I am saying them. Prior to saying those words, I am consciously aware, vaguely and wordlessly, of an idea that I want to convey. But the actual words, the message versus the idea, enter my consciousness later. I cannot, therefore, defend any claim that I consciously choose those words. When the AI behind Siri, Alexa, and Google Home speak to me, do they consciously choose their words?

MY BRAIN'S MOUTHPIECE

Some of my best ideas have come to me while speaking. The words that come out of my mouth surprise even me by expressing a new idea. I did not have that idea in my head until I put it into words. The process of synthesizing words from vague ideas stimulates thought and, apparently, creativity. I have always told my students that creative research requires writing, speaking, and collaborating because all three stimulate ideas. The mere act of trying to explain to someone else what you are thinking changes what you are thinking.

Some thinkers have asserted that thoughts are actually *nothing but* words and language. In *The Will to Power*, as translated by Walter Kaufmann, Friedrich Nietzsche wrote,

We think only in the form of language. … We cease to think when we refuse to do so under the constraint of language.[2]

This text, dated 1886–1887 in *The Will to Power*, has been poetically misquoted in many places as, "We have to cease to think if we refuse to do it in the prison house of language." Although this metaphor of a prison house of language has captured considerable currency with postmodernists, deconstructionists, and social scientists, the attribution to Nietzsche is questionable. Even the text, *The Will to Power*, is questionable, as it was published posthumously by Nietzsche's sister and perhaps misrepresented as his culminating work, when it may in fact just have been a random collection of notes. Nevertheless, Nietzsche has come to be associated with a strong connection between thought and language.

Following Nietzsche, Ludwig Wittgenstein later wrote,

The limits of my language mean the limits of my world. (Die Grenzen meiner Sprache bedeuten die Grenzen meiner Welt.)[3]

Steven Pinker, in his 2002 book *The Blank Slate: The Modern Denial of Human Nature*, sharply criticizes the view that thought and language are so inextricably intertwined. He argues that this belief is a consequence of the "blank slate" or "*tabula rasa*" hypothesis that denies human nature and assumes that all human behavior and thought is learned.

Based on my own introspective experience, I have to agree with Pinker. I certainly have thoughts before I have put them to words, and the fact that the words I say do not always match those thoughts speaks more to the limitations of language than to the limitations of thought. That the words can stimulate new thoughts speaks to the power of noisy feedback, which I will argue is at the heart of creativity. The British writer G. K. Chesterton, in 1904, expressed my position beautifully:

Every time one man says to another, "Tell us plainly what you mean?," he is assuming the infallibility of language: that is to say, he is assuming that there is a perfect scheme of verbal expression for all the internal moods and meanings of men.... He knows that there are in the soul tints more bewildering, more numberless, and more nameless than the colours of an autumn forest.... Yet he seriously believes that these things can every one of them, in all their tones and semi-tones, in all their blends and unions, be accurately represented by an arbitrary system of grunts and squeals.[4]

This is the nature of words. They are imperfect representations of thought, and yet they are simultaneously the scaffolding of thought. It is quite difficult to firmly grasp a thought in my own head without putting words to it, but once I've put words to it, I've likely sacrificed some of the essence of the thought.

Normal speech, like what I usually use when speaking publicly, does not, apparently, arise from my conscious mind. Conscious awareness of the words I say follows rather than leads. Yet I intended to say them (most of the time). So where do the words come from? Clearly, they come from the physical processes in my brain, and while they seem essential to my consciousness, which arises from the same physical processes, the words are not the result of conscious thinking.

FREUDIAN SLIP

Back in the 1890s, Sigmund Freud's focus on the importance of unconscious mental processes caused quite a furor because, at that time, most thinkers were die-hard Cartesian dualists, separating mind from matter.

To a dualist, the brain is just the mechanism that translates the intentions of an immaterial mind into muscle actions that produce speech. But to me, the brain is the hardware that hosts the processes of thinking, both conscious and unconscious.

What about the words on this page? Did they arise from my conscious mind? I can't really know how other writers work, but I can tell you how I work. I first form words in my head, as if speaking them, and then command my fingers to type out those words. The words that form in my head, just like natural speech, sometimes surprise me, and I become consciously aware of them only after they have formed. I then reread what I wrote and almost always revise it. That's not really what I meant, or those words are jumbled and my readers won't understand what I meant. I can't do that kind of back-up-and-revise when speaking out loud, but I can when writing. And yet, when writing, I *am* speaking out loud, or at least, hearing in my head the words I would have said were I speaking out loud.

One of the truly remarkable properties of the human brain is that it can synthesize signals within the head that match the signals that would be generated by our sense organs. Since at least the 1800s, psychologists have studied the phenomenon that the brain can internally synthesize stimulus that would result from sensing some action commanded by the brain. This internal feedback signal is called an "efference copy." I will have more to say about efference copies in chapter 5, where I consider the question of whether digital machines have similar mechanisms.

My ears sense sound and trigger patterns of neuron firings that, within my brain, turn into the sensation of comprehension of spoken language. But even without help from my ears, other parts of my brain can cause the same patterns of neuron firings that fool my brain into thinking it has heard those words out loud. The way that I write is that my brain puts my vague thoughts into words, unconsciously, and my brain synthesizes voices in my head that make me consciously aware of those words, as if I were hearing myself speaking them.

The brain is capable of synthesizing both fake sensing and fake actuation. My brain can produce sequences of words as if I were to speak them, but bypasses my mouth and ears and feeds those words right back to the language centers as if I had spoken them and heard myself speaking. We will see that some digital artifacts also have feedback mechanisms, but they do not yet rise to the same level of sophistication.

MONKEY MIND CONTROL

The ability to bypass the actual physical sensing and actuation, it turns out, is a property of at least primate brains more generally. My colleague at Berkeley, José Carmena, while he was a postdoc working with Miguel Nicolelis in the Department of Neurobiology at Duke University, conducted a series of experiments that showed that monkeys could learn to control a virtual world through their thoughts alone.[5] Specifically, monkeys were instrumented with cortical implants that could sense neural activity. The monkeys were then taught to control icons on a screen using a joystick. They would learn that with the joystick, they could steer one icon toward another one and get a reward when the two icons merged. The cortical implants were used to record the neural activity that was driving the monkey's muscles to move the joystick.

A motor signal from the central nervous system to the muscles on the periphery is called an "efference." After recording these efferences for a while, the researchers learned to associate patterns of neural activity with the muscle movements that moved the joystick. They then disconnected the joystick and drove the icons directly from the neural signals. Then, a remarkable thing happened. Quite quickly, the monkeys realized that they no longer needed to physically move the joystick, so they stopped moving it. Instead, they "thought" the icons together and got the reward.

I have always been somewhat averse to exercise. When I was in high school, I discovered that I could improve my tennis game by just lying in bed at night and imagining that I was playing tennis. I would practice my serves and feel them go into the net or out of bounds. I would feel catching the sweet spot where the racket engages with the ball in perfect balance. After a few imaginings, my serves got better, and sure enough, the next day, my real serves would be better too. Like Carmena's monkeys, I could "think" rewards into reality.

I didn't consciously plan to add that tennis anecdote to the story here. It popped into my head as I was writing about Carmena's work. My brain reacted to "hearing" about the monkeys by forming a vague connection between monkeys and a personal experience more than forty years ago. And then, the thought became words, all without making or hearing a sound. Thoughts arise when my conscious mind becomes aware of the words articulating an earlier thought. And each thought turns into words in a synthetic feedback loop that mimics speaking the words and hearing myself speak them. This brings me, finally, to the essential question of this chapter: Can digital computers do that? Is this feedback theory of mind accessible to digital beings?

FROM GOFAI TO MACHINE LEARNING

The most dramatic recent progress in AI comes directly from the introduction of feedback. Good old-fashioned AI, known as GOFAI to its friends,[6] has no such feedback. In the GOFAI style of AI, which dominated the field from the 1950s through the 1980s, knowledge is represented symbolically in a human-readable form, and laboriously constructed data structures represent the relationships between symbols. More recently, AI has come to be dominated by statistical techniques and optimization algorithms, including artificial neural nets, where representations of knowledge are learned from data rather than being provided by humans. Some people call this more recent approach "computational intelligence."

An early example of GOFAI is the ELIZA program, built in the 1960s by the MIT computer scientist Joseph Weizenbaum. ELIZA was capable of engaging humans in a conversation by responding in a manner bearing a remarkable resemblance to an empathic psychologist. Using a modern online implementation of ELIZA called "elizabot," I had the following short conversation with ELIZA:[7]

ELIZA: Please tell me what's been bothering you.

ME: I think my computer is smarter than me.

ELIZA: Do computers worry you?

ME: Yes.

ELIZA: You are sure.

ME: No, not really.

ELIZA: Are you saying no just to be negative?

Weizenbaum claimed that ELIZA demonstrates that the *appearance* of human-like intelligence does not in fact imply the *existence* of humanlike intelligence:

Machines are made to behave in wondrous ways, often sufficient to dazzle even the most experienced observer. But once a particular program is unmasked, once its inner workings are explained in language sufficiently plain to induce understanding, its magic crumbles away; it stands revealed as a mere collection of procedures, each quite comprehensible. The observer says to himself, "I could have written that." With that thought he moves the program in question from the shelf marked "intelligent" to that reserved for curios, fit to be discussed only with people less enlightened than he.

Weizenbaum's goal with ELIZA was to demonstrate that the appearance of intelligence was easy to achieve and therefore should not be construed as evidence of real intelligence:

The object of this paper is to cause just such a reevaluation of the program about to be "explained." Few programs ever needed it more.[8]

An aspect of Weizenbaum's statement is disturbing. He seems to be claiming that if a program is comprehensible, then it must not be intelligent. This implies that we will never understand intelligence because once we understand it, we will no longer consider it to be intelligence.

GOFAI led to the creation of so-called "expert systems," which encode the experiences of human experts in large sets of production rules, which are essentially if-then rules. If you see this text, respond with that text. The brothers Hubert and Stuart Dreyfus, both Berkeley professors, Hubert in philosophy and Stuart in engineering, sharply criticized the concept of expert systems in their 1986 book, *Mind Over Machine*. They pointed out, quite simply, that following explicit rules is what novices do, not what experts do.

Think about learning a new language. When you are a novice, you explicitly apply the rules of grammar and look up (in your head) words that are translations of words in your native tongue. When you become an expert, you do not follow rules, at least not consciously. You just speak.

Dreyfus and Dreyfus slammed the hype that surrounded AI at the time, calling its high priests,

false prophets blinded by Socratic assumptions and personal ambition—while Euthyphro, the expert on piety, who kept giving Socrates examples instead of rules, turns out to have been a true prophet after all.[9]

Here, Dreyfus and Dreyfus are reacting (rather strongly) to what really was excessive hyperbole about AI at the time. They were just the tip of a broad backlash against AI that came to be called the "AI winter," where funding for research and commercial AI vanished nearly overnight and did not recover until around 2010.

Computational intelligence, as it evolved much later, works primarily from examples, "training data," rather than rules. The explosion of data that became available as everything went online catalyzed the resurgence of statistical and optimization algorithms that had been originally developed in the 1960s through 1980s but lay dormant through the AI winter before exploding onto the scene around 2010.

SMILING CATS

To me, the most interesting aspect of the computational intelligence algorithms is their use of feedback. The principle of feedback, which I will elaborate in chapter 5, is that the output is used to adjust what the system does in the future. When my brain produces a sentence, and I then become consciously aware of the words I have used, that conscious awareness affects the subsequent thoughts that form, which then affects the subsequent words I will use. This is a classic feedback system.

The explosive renaissance of AI around 2010 was based on a family of algorithms with a long history. Although the algorithms had been known for several decades, researchers suddenly started showing startling success on hard classification problems such as recognizing things in images. Even these same algorithms had not done very well on these same problems before the AI winter, probably because there just wasn't enough available data to train them properly. Amazing improvements started appearing on many fronts at once. Image analysis, for example, became good enough that you could reliably search the Internet for pictures of smiling cats (see figure 4.1). Facebook could begin automatically tagging recognized people in uploaded photos. Handwriting recognition became good enough that banks started allowing you to deposit checks by taking pictures of them with your smartphone. Voice recognition became good enough to enable smart speakers, a whole new category of

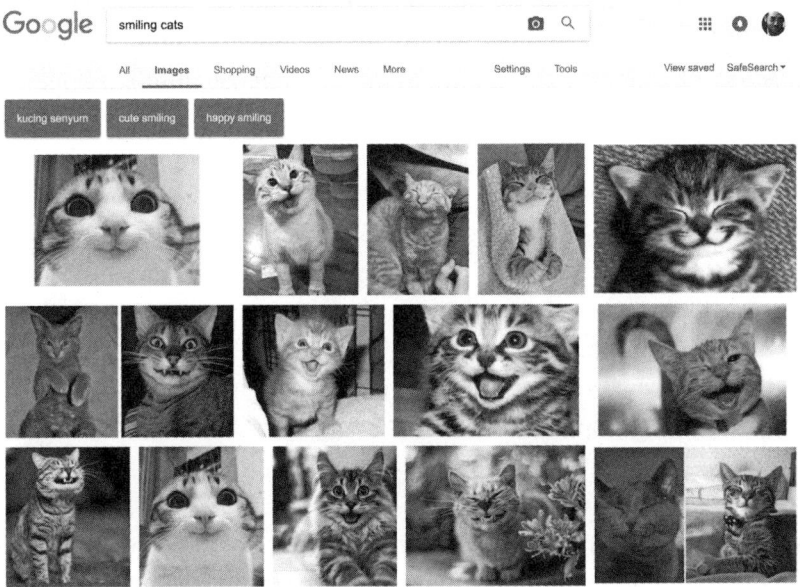

4.1 Results of a Google search for "smiling cats" (retrieved June 18, 2018).

consumer product. And machine translation got good enough that I can now reliably read a PhD thesis that was written in German, although I speak no German.

The algorithms behind this renaissance are nothing like the production rules of GOFAI. A central one of these algorithms, now called back-propagation, first showed up in automatic control problems quite some time ago. In 1960, Henry J. Kelley, an engineer at Grumman Aircraft Engineering on Long Island in New York, traveled to Los Angeles to attend the semiannual meeting of the American Rocket Society and present a paper on how to synthesize a controller that would carry a spacecraft from Earth's orbit to Mars's orbit around the sun using a solar sail.[10] To control the spacecraft such that it follows an optimal path from one orbit to another, a mechanism would vary the angle of the solar sail. The problem that Kelley addressed was how to determine what angles to use. The procedure he derived is essentially the same as backpropagation, although his formulation is more continuous than the discrete form used in machine learning today. Based in part on Kelley's work, in 1961,

Arthur E. Bryson and some colleagues at Raytheon's Missile Systems Division in Rhode Island were working on the problem of how to control a spacecraft that is re-entering the Earth's atmosphere to minimize heating due to friction. Depending on the vehicle, they can either control the lift (as in a glider) or the drag (for vehicles without wings). To solve this problem, they adapted Kelley's method into a multistage algorithm that closely resembles the backpropagation technique used for deep neural nets today.[11] So the techniques most widely used today in machine learning were literally first developed by rocket scientists!

The Kelley-Bryson technique was restated in a form closer to its usage today in 1962 by the same Stuart Dreyfus who accused the AI researchers of being "false prophets," who, at the time, was working at the RAND Corporation in Santa Monica, California.

Kelley, Bryson, and Dreyfus saw these algorithms as techniques for solving an optimal control problem. They no doubt would have been very surprised to see the algorithms applied to classifying images as "smiling cats" versus "frowning cats" (see figure 4.2). In their modern usage, these algorithms are used to learn how to classify data.

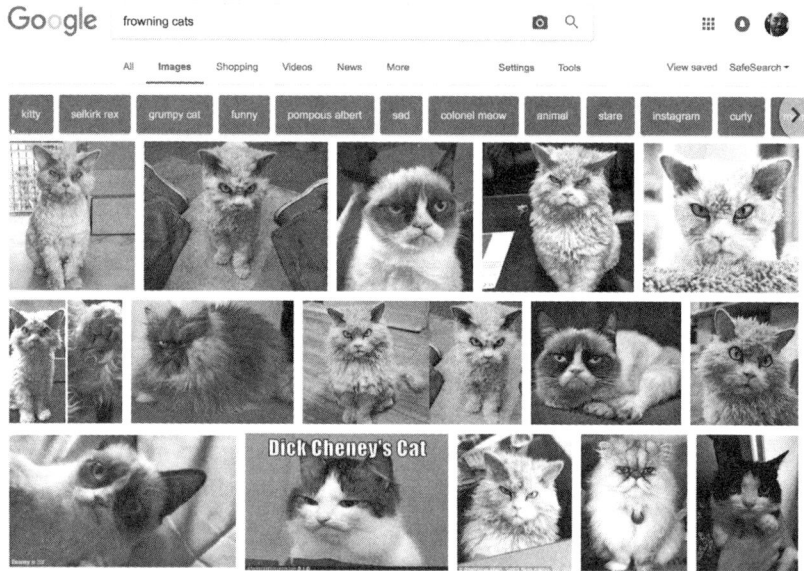

4.2 Results of a Google search for "frowning cats" (retrieved June 18, 2018).

Roughly, the algorithms work as follows. First, an image or a sound is digitized and represented as a collection of numbers. Then, a computation is set up that takes as input these numbers and outputs a classification. For example, the function, when computed on an image, may declare its output to be one (smiling cat) or zero (frowning cat). For simplicity, I am assuming that all images are of cats and that every cat is either smiling or frowning. In practice, the way the algorithms work today is that they generate not just zero or one, but rather a number between zero and one where "close to one" means "likely to be a smiling cat," and "close to zero" means "likely to be a frowning cat."

Initially, of course, we have no idea how to set up the computation to reliably give us a classification. So, at first, the algorithm will do very badly, generating essentially random classifications. The secret is to train it until it gets better. The strategy used in the approach that is now called "deep learning" uses a particular structure for this computation that is roughly inspired by what neurons do, and consequently, the computation is called a neural net or artificial neural net. The "deep" in "deep learning" refers to a layering of computations, where one layer of artificial neurons feeds data to the next layer of artificial neurons until, at the final layer, out comes the classification, a number between 0 and 1 for our example.

LEARNING AND FEEDBACK

In a neural net, there are many parameters to be chosen. A layer with thousands of neurons has a huge number of parameters, numbers which, if set correctly, will result in a correct classification. But we have no idea what those numbers should be. So what people do is astonishingly simple. Initially, the neural net is set up with randomly chosen parameters, and consequently, of course, it does a terrible job of classification. Give it an image of a smiling cat, and it may produce, for example, the number 0.42. Now we apply feedback. We assume that the image is correctly labeled as a smiling cat, so the computer knows the right answer it should have given, namely 1. But its computation yielded 0.42, which is too low. It now uses the Kelley-Bryson-Dreyfus algorithm to adjust its parameters a bit so that, given the same image, its answer will be a bit closer to the correct answer, say 0.48. Now give it another image, this time a "frowning cat," and it gives another answer that is probably incorrect. Again adjust the parameters a little bit so that the answer, for the same image, will become a bit closer to the right answer. Keep doing this on a large number of images, and the neural net starts giving very good answers for images it has never seen before.

It is worth pointing out a few things here. First, this procedure requires a great deal of computation. An image taken on my (somewhat old) smartphone has eight million pixels, each of which is represented by three numbers. Hence, the input to the algorithm at the start is twenty-four million numbers for each image. Effective neural nets even for relatively simple problems can have a large number of parameters, and these parameters are combined with the numbers from the image by multiplying, adding, and exponentiating. There are many tricks we can perform that will reduce the number of numbers that have to be crunched, and you can attend many technical conferences where researchers compare notes to determine which techniques work best, but all known techniques require a serious commitment of computer resources. The 2010 renaissance in the field is due in no small part to the availability of inexpensive computers with an astonishing ability to crunch numbers.

Second, these algorithms require a great deal of training data. After one or two images, the parameters may be slightly closer to yielding correct classifications, but to get really good, they need millions of labeled images. The 2010 renaissance, therefore, was also fueled by the availability of large numbers of images. Why do you think Google, Amazon, Facebook, and Apple are all willing to store your photos and videos online for free? These images have value because they can be used to train neural nets.

The algorithm I just sketched for you is one of a class called "supervised learning" algorithms. It is "supervised" in the sense that it is given the right answers, labels for each of the images it trains on. In practice, getting images labeled can become the bottleneck. Fortunately, people on the Internet tend to label images, which can help. And people can be induced to provide labels, for example in Facebook by tagging your friends. Finally, people can be paid small amounts for each classification, using for example Amazon Mechanical Turk.

PERCEPTRONS

The psychologist Frank Rosenblatt (1928–1971) was perhaps the first to explore the potential of artificial neural nets. In the 1950s he designed an electronic device that he called a "perceptron" that was loosely inspired by biological neurons. Using simulations on an IBM 704 computer at the Cornell Aeronautical Laboratory, he showed that perceptrons could learn to classify images of simple geometric shapes.

As with the modern techniques, Rosenblatt used a labeled training set and feedback to learn the parameters of each perceptron that would lead to reliable classification. Rosenblatt, therefore, may be the first person to show an algorithm that could learn from examples. Unlike the modern techniques, he used only a single layer, and his perceptron output was binary rather than numeric.

Rosenblatt's work generated quite a lot of excitement, publicity, and controversy, culminating in a 1969 book entitled *Perceptrons* by Marvin Minsky and Seymour Papert from MIT. Minsky and Papert were sharply critical of Rosenblatt's formulation, focusing on the limitations of perceptrons. They were either oblivious to or ignored the potential improvements that derive from using multiple layers. Their book threw a wet blanket on the whole field, which then experienced its own mini AI winter until interest resurged in the 1980s,[12] only to sputter out again in the 1990s and 2000s.

Although I didn't realize it at the time, in 1980, I was using learning algorithms strikingly similar to Rosenblatt's perceptrons, although with a structure more closely resembling what are called today "convolutional neural nets." I was working at Bell Labs in New Jersey on the design of voiceband data modems, devices that could send bit sequences over ordinary telephone lines. These devices were important to the early development of the Internet and became an essential peripheral for every personal computer in the early 1980s. One problem with telephone lines is that they distort the signal. The distortion, by design, does not interfere with the intelligibility of a voice signal, but it does interfere with the ability to determine at the receiving end what bits were sent at the transmitting end. When first establishing a connection, a voiceband data modem sends a training sequence that is known a priori by the receiving modem, and the receiving modem uses a learning algorithm to set the parameters of an "adaptive equalizer" that learns the distortion of the channel and reverses it. Once the equalizer has learned to compensate for the channel, the transmitting modem switches to sending bits that are not known a priori to the receiving modem. Since the receiving modem has learned to effectively compensate for the distortion, it reliably decodes these bits. Interestingly, it also makes the assumption that it has correctly decoded the bits, and therefore, under this assumption, it can continue to run its learning algorithm as it did with the training sequence. This means that the distortion in the channel can change over time, and as long as the change is slow enough that bits continue to be reliably decoded, the receiving modem continues to learn and compensate for the changing distortion. I subsequently co-authored a textbook about these data communication techniques.[13] The algorithms were strikingly similar to those used today in machine learning.

The learning algorithm used by Rosenblatt did indeed have quite a few limitations, so Minsky and Papert's criticisms had some validity. Most important, Rosenblatt's algorithm was not immediately adaptable to multiple layers of artificial neurons. This required the Kelley-Bryson backpropagation algorithm, which was first applied to artificial neural nets in 1986 by David Rumerlhart, Geoffrey Hinton, and Ronald Williams, who were apparently unaware of the earlier work of Kelley, Bryson, and Dreyfus (Hinton won the 2019 Turing Award for this work).[14] Their innovation was key to the 1980s resurgence of interest in artificial neural nets.

Compared to GOFAI, the essential innovation of artificial neural nets is the introduction of feedback. The network produces an output, measures the degree to which that output is wrong, and revises its parameters so that it will do better the next time.

FROM JELLYFISH TO DOGS

A startling recent development in neural nets research underscores the power of feedback by adding another higher-level feedback mechanism above backpropagation. A technique called "generative adversarial networks" (GANs) has two machines learn from each other, pitting one against the other in a tight feedback loop. *MIT Technology Review* calls Ian

Goodfellow, a researcher on the Google Brain project, the "GANfather" for his invention in 2014 of GANs.[15] According to *MIT Technology Review*, Goodfellow came upon this idea while he was a PhD student in Montreal when he went out drinking to *Les 3 Brasseurs* (The Three Brewers) to celebrate with fellow doctoral students. His colleagues asked for help with a project they were working on to get computers to create synthetic but realistic photos on their own.

Researchers had already figured out how to use neural nets to synthesize images, but the images were not very realistic. In fact, many were positively bizarre. The Google DeepDream program, created by Alexander Mordvintse, illustrates one approach to creating synthetic images. Once a neural net has been trained to recognize dogs, for example, then the backpropagation algorithm can be modified so that, given a new input image, instead of adjusting the parameters of the neural net, the algorithm adjusts the input image. The goal is to synthesize a variant of the input image to which the neural net will exhibit a strong response of "dogness." An example of the results of this algorithm is shown in figure 4.3. The original image, shown at the top, is an underwater photograph of jellyfish. After ten iterations of Mordvintse's program, the neural net modifies the image so that the jellyfish begin to acquire snouts, legs, and tails in a weird arrangement. After fifty iterations, the result (shown at the bottom in the figure) is positively hallucinogenic, prompting some researchers to speculate that drug-induced hallucinations result from similar reversals of the visual system in the human brain.

Ian Goodfellow's idea was to play off two neural nets against one another. One of the networks, called the discriminative network, would have the goal of classifying images as either synthetic or natural, just as we might train a network to classify images as either smiling or frowning cats. The second network, called the generative network, will produce synthetic images with the goal of increasing the error rate of the discriminative network. That is, the generative network tries to fool the discriminative network. Because the discriminative network is trying to *reduce* its error rate, this is a kind of adversarial game, hence the name "generative adversarial networks." The two networks are in a tight feedback relationship with one another.

4.3 Synthetic images created from the top image by Google's DeepDream. By Martin-Thoma, public domain, from Wikimedia Commons.

4.3 (continued)

FEEDBACK IN BIOLOGY

Feedback is also essential to biological beings, even at the lowest levels, well below cognition. If a bacterium moves in a direction where the density of nutrients is decreasing, its sensors will detect the error and reverse its direction of travel. At a much higher level, converting thoughts into speech in a human brain involves synthesizing motor signals—efferences—and feeding these signals back into the brain, which synthesizes anticipated auditory consequences of these motor actions. The brain will detect any errors, such as I meant to say this but said that, and correct those errors. The brain is also using such feedback at a much lower level to adjust the motor signals at a very fine-grain level so that the sounds come out as expected. This is why many hearing-impaired people have difficulty speaking clearly. The feedback loop is essential, and the hearing impairment breaks the loop.

So how much feedback is there today in digital technology? Actually, remarkably little, primarily because computers have very limited sensing and actuation capabilities compared to humans. But this is changing rapidly. The 2010 renaissance of AI was driven in part by vast amounts of data going online. What will happen as all our *things*, not just our data, go online? Computers are rapidly acquiring hugely greater visibility into the physical world, and hugely greater ability to affect the physical world. This could lead to a renaissance that will make today's AI renaissance look like the dark ages.

In the next chapter, we dive more deeply into what feedback really is, how it works, and how it can be layered. A key observation is that feedback is not necessarily an iterative, algorithmic process, as some readers may assume.

5

NEGATIVE FEEDBACK

TALKING TO MYSELF

Occasionally, the talks that I give get recorded. I can't stand watching those recordings, particularly listening to my voice. Is that really my voice? Who is this annoying guy? While I know it is me speaking, it doesn't feel like me. Why is the sensation of listening to a recording of myself so different from the experience of speaking? When I speak, my ears sense the sounds I make as I speak, but my brain does not hear myself speak the same way it hears other people speak. If I am listening to a recording, however, it feels like someone else is speaking.

Part of the reason for this is that speech-caused vibrations propagate through the skull and tissues to my inner ears, causing a distinctly different stimulus from that caused by sound waves coming from outside my head. But there is a deeper reason. Even the most primitive biological nervous systems separate self from non-self; they distinguish sensory stimulus that is caused by their own actions from that caused by something in the environment. Your brain changes what your ears hear to handle your own speech differently from external sounds. It turns out that some digital systems already have rudimentary forms of this reflexive self-awareness. Many more are likely to have this capability in the future.

Recall that a signal from the central nervous system to the muscles on the periphery is called an "efference" or an "efferent signal." When you speak, your brain issues efferent signals to the muscles in your chest, larynx, mouth, and tongue. Your muscles then shape the acoustic chamber that gives each phoneme its characteristic sound. Your brain, however, cannot do this very precisely without help from your ears. Assuming you are not deaf, your ears participate directly in the production of speech sounds. One part of your brain, that commanding the muscles, tells another part, that interpreting sound, what it should expect to hear.[1] If the sound does not quite match the expectation, your brain will make small corrections to the efferent signal to get the sound to match the expectation better. It does this continually, while speaking, quickly enough that there is no perceived delay. Your speech sounds are made to match your expected sounds. Such a mechanism is a classic example of what engineers call "negative feedback control."

Figure 5.1 shows a diagram of how this likely works. More accurately, the diagram shows how I, an engineer, would design such a system, were I charged with doing that. I suspect the real system in our bodies is much more complex and does not divide so nicely into discrete components connected by thin wires. Nevertheless, we can use this diagram to understand negative feedback better and establish some relationships between how computers work, something we understand fairly well, and how biological beings work, something we are rapidly learning about but are largely still in the dark.

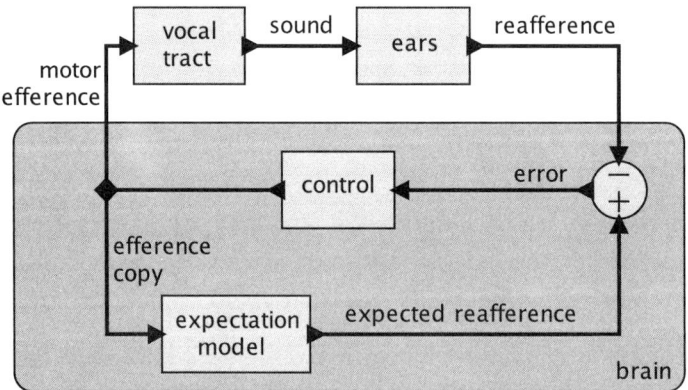

5.1 Speech production by means of negative feedback.

In the center of the figure is a mysterious box labeled "control." This represents the brain functions that decide what to say and produces the commands, labeled "motor efference" in the diagram, sent to the muscles to cause the body to make sound. These signals drive the vocal tract to produce sound that is then perceived by the ears, generating sensory stimulus back to the brain that is called "reafference." The term "reafference" refers to sensory signals that are generated by an animal's own actions, distinguished from "exafference," which are sensory signals that result from external stimuli from the environment.

While the body is producing sounds that the ears are picking up, at the same time, the brain generates an "efference copy," according to this theory, which is fed back into a different part of the brain that calculates what the ears should be hearing, an "expected reafference." How the brain does this is a bit of a mystery. The mystery is represented in the figure by a box labeled "expectation model" that converts the efference copy into the expected reafference. I suspect that this expectation model is learned, perhaps by a mechanism similar to the backpropagation algorithm that we saw in the previous chapter.

Now comes the most important part of this whole operation. The brain compares the reafference, what the ears are hearing, with the expected reafference, what it expects the ears to be hearing. This comparison is shown in the figure by a small circle that suggests that one signal is simply subtracted from the other to get an error signal. If the two signals are identical, the error is zero, so the ears are hearing exactly what is expected and no change is required in the control. If what the ears are hearing does not match the expectation, the error is nonzero, and the brain will try again to make it zero by modifying the signals it sends to the muscles.

SPEAKING LOUDLY ALL AT THE SAME TIME

The error signal, the difference between what the brain expects to hear and what it actually hears, has two important purposes. First, it can be used by the brain to correct the efference, the signals to the muscles, to make the sounds you make better match the sounds you expect. Without this feedback, you would not speak very well. This is why hearing-impaired people often have difficulty pronouncing words intelligibly. They can't sense their own distortions.

Second, the error signal necessarily contains whatever the ears are hearing that the brain does *not* expect. The brain is not able to cancel out what it does not expect. As a consequence, this feedback mechanism gives the brain a way to distinguish self from not self. Sounds that do not come from oneself and are not expected cannot be cancelled. This could explain why you are often able to hear and understand what another person is saying even if you are speaking at the same time.

For the first of these purposes, consider a simple aspect of speech production, namely speaking at an appropriate volume. We have all probably had the experience of talking over loud music, when suddenly the music stops. The speech that comes out of our mouths then is much too loud, maybe embarrassingly so, even disastrously so, depending on what we were talking about. "Joe over there is a real jerk!," you say for everyone in the room to hear. Your brain very quickly realizes the error and adjusts the loudness of your voice.

With this loudness example, we can understand the term "negative feedback." Why is it negative? In the figure, the reafference, what the ears hear, is subtracted from the expected reafference. If you are speaking too loudly, the reafference will be larger than the expectation, and consequently the error will be negative. The control interprets this negative error signal as an indication that it should reduce the volume. If we had miswired the diagram and instead subtracted the expectation from the reafference, then the sign of the error signal would be wrong, and when we perceive ourselves talking too loudly, we would simply talk even louder. That style of control is called "positive feedback."

FEEDBACK FROM BELL LABS

The use of the term "negative" in connection with feedback appeared in the 1920s at Bell Labs, where the electrical engineer Harold Stephen Black (1898–1983) was working on vacuum tube amplifiers for transatlantic telephony. One day, while traveling from New Jersey to New York by ferry, he got an inspiration for how to dramatically improve the quality of these amplifiers. With nothing to write on but his *New York Times*, he wrote his notes in the margins and signed and dated it to establish a record of the time of the invention he would subsequently patent.

Black observed that, at the time, vacuum tube amplifiers had gotten good enough that it was relatively easy to achieve a very high gain, taking a small, weak electrical signal in and producing an output signal with millions of times more energy. But the output signal would be distorted compared to the input, like the speech of a hearing-impaired person. He realized that by using negative feedback, he could sacrifice some gain in exchange for a much more faithful output signal. In a famous 1934 paper, he reported that by sacrificing a factor of ten thousand in the gain, he could reduce the distortion by a factor of ten thousand.[2] This technique was so effective that, henceforth, the term "open loop" became a derogatory term applied to any system without feedback.

The use of the term "negative" makes sense in Black's application because the error signal was a simple scalar, a measurement of the amplitude of an electrical signal. Consequently, the error signal, also a scalar, could only be positive, negative, or zero. In more complicated uses of negative feedback, such as backpropagation and speech production, the error signal is not so simple, so better terms might be "self-correcting," "balancing," or "discrepancy-reducing" feedback. But the term "negative feedback" is so established in engineering circles that it is hard to avoid.

POSITIVE FEEDBACK

Positive feedback is where the correction is made in the wrong direction. Most of you, dear readers, maybe excluding the AIs among you, have experienced positive feedback. It can be quite unpleasant, despite the positive connotation of the term. Specifically, in a room with a microphone and speakers, such as a public address system, the goal of the system is to make sounds picked up by the microphone louder. If you put the microphone in front of the speaker, you can get positive feedback that results in very loud squeal coming out of the speakers. Why?

The speakers react to a signal from an amplifier, analogous to the efferent signals from the brain to the muscles, by producing sound. That sound fills the room, and the microphone picks it up, sending the resulting reafference back to the sound system. If the sound system is too dumb

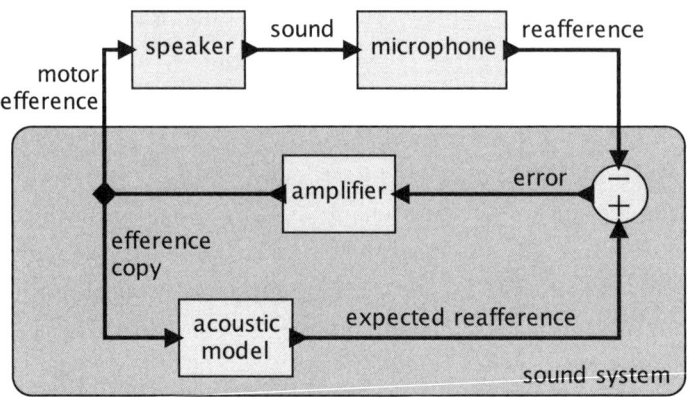

5.2 Audio production with echo cancellation.

to realize that the sound it just picked up is the one it just produced, then it will simply amplify the sound again and send it back to the speakers. The microphone again picks up this louder sound and amplifies it further, making it even louder. The result is an unstable system that will keep increasing the volume of the sound until it can do so no more. Either protection circuitry in the amplifier will limit the volume at some maximum, a fuse will blow, or the speakers will blow. All of the outcomes are unpleasant.

Fortunately, many sound systems today are much better than this. They do, in fact, include an additional loop where an efference copy is fed back into an expectation model, as shown in figure 5.2. The technology is called "echo cancellation." Note that the structure of the blocks in that figure is identical to that of figure 5.1. Only the labels on the boxes have changed.

Echo cancellation is not trivial to realize because the sound system has to build an expectation model that turns the efference copy into an expected efference. But the sound propagates from the speaker to the microphone in a room, bouncing off the walls and being partially absorbed by the carpet. That acoustic path distorts the signal, so the shape of the signal picked up by the microphone will not perfectly match the shape of the signal sent to the speakers. The sound system needs to construct, on the fly, a model of the acoustic properties of the path from speakers to the microphone. This is exactly what echo cancellation does, and it uses an algorithm strikingly similar to back-propagation. The goal of the algorithm is to make the error signal as close as possible to zero.

Echo cancellation is used in smart speakers, such as the Amazon Echo or Google Home, to improve their ability to hear commands while they themselves are playing sound. They can be playing music quite loudly and still hear a human voice saying "Alexa" from across the room, this despite the fact that the microphone is right next to the speaker, so the sound it picks up from the speaker is many orders of magnitude louder than the distant human voice. Very good echo cancellation is essential.

Smart speakers also use a technique called "adaptive equalization" to accomplish something quite analogous to the first goal of the efference copy, making higher-quality sounds. When playing music, the smart speaker "knows" what the music should sound like. It should sound like what is in the MP3 file. But its speaker is far from perfect. It is housed in a small box, and without any feedback, the sound it would produce would really not be very good. It would likely sound tinny and distorted. But since the speaker knows what sound should appear at its microphone, it can measure any error and use it to adjust a filter that precompensates for the distortion of its speaker. This filter is called an adaptive equalizer. It is "adaptive" because it learns the distortion, again using an algorithm strikingly similar to backpropagation.

COGNITIVE FEEDBACK

Peter Godfrey-Smith, who appeared in chapter 2 with his study of octopuses, emphasizes that feedback is essential to sentience and other cognitive functions.[3] He points out that you need not just sense-to-act connections, which even bacteria have, but also act-to-sense. You have to affect the physical world and sense the changes. Sense-to-act is open loop; you sense, you react. Combine this with act-to-sense, and you close the loop.

Humans use feedback at higher cognitive levels in a closed loop. At the language level, given a sense of what we want to say, words come out of our mouth. Only then do we become consciously aware of those words, and only then can we determine whether they accurately reflect the sense of what we wanted to say. We will correct it if it came out wrong. And what we said will affect what we say next.

Amazon Echo and Google Home have none of this higher-level feedback, at least not as of this writing. To demonstrate this, it is easy to get them into an utterly silly conversation with each other. I say, "Alexa, tell me what the first thing on my calendar today is." The Amazon Echo responds,

The first thing on your calendar today, at 9 a.m., is "OK Google, what is the last thing on my calendar today?"

Google Home responds,

The last thing on your calendar today starts at 3 p.m., and the title is, "Alexa, what is the first thing on my calendar today?"

Alexa responds,

The first thing on your calendar today, at 9 a.m., is "OK Google, what is the last thing on my calendar today?"

Google Home then responds identically to its first response, and the two cheerily continue the conversation until today has expired and something else is on my calendar. They are oblivious to the trick I played on them, the meaning of their sounds, and the repetition of their actions. Part of their charm is that they are not in the least embarrassed by their stupidity.

SELF AND NON-SELF

Many psychologists today believe that efference copies are central to our sense of self. They distinguish self-induced from not self-induced sensory stimulus. All animals with sensors have evolved some form of efference copy mechanism because otherwise they would react to their own actions as if those actions were imposed by their environment. Do all animals have a sense of self? Almost certainly, a worm's sense of self does not resemble our own, so "sense of self" comes in degrees. When a human thinks, "yes, this is what I meant to say," she is operating at a very different level than when a worm senses that the ground is moving under it because it is contracting and expanding its body. But the essence of the

5.3 First-generation Amazon Echo and Google Home talking to each other in an infinite loop.

mechanism is the same, negative feedback, just operating on signals of vastly different complexity.

The importance of the efference copy has been understood at some level since the nineteenth century.[4] Johann Georg Steinbuch (1770–1818), working in Erlangen, Germany, published a book that illustrated the essential concept with a simple experiment. He noticed that if you hold your hand still and roll an object around in it, say, a spoon, you will not be able to recognize the object from the sensations coming from your hand. But if you actively grasp and manipulate the object, you will quickly recognize it as a spoon. The motor efference, therefore, must play a role in recognition, which implies that the motor efference must be fed back to the sensory system.

Later in the nineteenth century, a German physician and physicist, Hermann von Helmholtz, observed that an efference copy must be needed to prevent the brain from thinking that the world is moving when the eye moves. The experiment he proposed is to gently press on your own eye. I tried this, and I don't recommend it. It is rather unpleasant to press on your own eye. But indeed, the world seems to jump. Interestingly, it does so even though it is ultimately my own motor action, via my hand, that is moving my eyeball. But this is not a usual connection between motor action and reafference, so my brain does not expect it. It has not learned that connection. I do not want to try this experiment, but I suspect that if I pushed my eyeballs enough, eventually my brain would learn to predict the sensory consequences and I would no longer perceive the world moving.

GUNS AND FEMURS

Feedback, it turns out, appears in biological systems in many ways, even some that have nothing to do with cognition or perception. For example, feedback is essential to the correct formation of a human body from a fetus. In effect, as a body grows, it figures out on the fly how to build itself using feedback mechanisms.

A naive view of genetics holds that the human genome encodes a complete description of a fully formed human body. All that biology has to do to build a body is realize the description encoded in the DNA, as if the DNA were a blueprint, and behold, a human body forms. But this cannot possibly be true. There is simply not enough information in DNA molecules. The complicated brain structure revealed by Jeff Lichtman and his team at Harvard, whom we encountered in chapter 3, proves this point. There is not enough information in the genome to encode this complexity. The structure, therefore, has to be learned during development, and feedback is an essential part of this process.

This same statement applies to much simpler biological structures than the brain. The psychologist Steven Pinker, who has written extensively about the effects of genetics versus environment, illustrates this nicely:

The genes that build a femur cannot specify the exact shape of the ball on top, because the ball has to articulate with the socket in the pelvis, which is shaped by other genes, nutrition, age, and chance. So the ball and the socket adjust their shapes as they rotate against each other while the baby kicks in the womb. (We know this because experimental animals that are paralyzed while they develop end up with grossly deformed joints.) Similarly, the genes shaping the lens of the growing eye cannot know how far back the retina is going to be or vice versa. So the brain of the baby is equipped with a feedback loop that uses signals about the sharpness of the image on the retina to slow down or speed up the physical growth of the eyeball.[5]

A pioneer in the use of use of feedback, the American mathematician Norbert Wiener developed technology for the automatic aiming and firing of anti-aircraft guns during World War II (see figure 5.4). Wiener had a huge impact on our understanding of feedback control systems. Wiener coined the term "cybernetics" for the conjunction of physical processes, computation that governs the actions of those physical processes, and communication between the parts. He derived the term from the Greek word for helmsman, governor, pilot, or rudder. The metaphor is apt for control systems, since a helmsman constantly uses negative feedback to

5.4 120-mm anti-aircraft gun, deployed near the end of World War II. These guns were automatically aimed using negative feedback by the M10 director system, which used information from a radar system, and the M4 gun computer. This gun is on display at the Washington National Guard Museum, located on Camp Murray in Washington State. Courtesy of the Washington National Guard State Historical Society.

adjust the path of a boat. When the boat veers to port, steer to starboard, and vice versa. Wiener realized that the mechanisms of negative feedback applied in both biology and technology. He entitled his pioneering paper on the subject, "Cybernetics: Or Control and Communication in the Animal and the Machine." He identified negative feedback as the case when:

The information fed back to the control center tends to oppose the departure of the controlled from the controlling quantity.[6]

Wiener and his colleagues had earlier noticed that the very notion of "purpose," having a goal, necessarily involves some form of negative feedback because to have a purpose, one has to be able to compare what has been accomplished against that purpose.[7] The aspirational nature of "purpose" implies that if the comparison reveals that the purpose has not been achieved, then one will adjust one's actions to come closer to achieving that purpose. This is the essence of negative feedback.

DELAYED FEEDBACK

A naive understanding of negative feedback is that it is an iterative process, where, in a sequence of steps, an output is produced, an effect is observed, the effect is compared against a desired effect, and a new output is constructed. But this is far too discrete, too algorithmic, and too backward looking. Black's amplifiers had no such steps. Neither did Wiener's anti-aircraft guns. Neither does femur formation nor speech sound production. A non–hearing-impaired person does not produce erroneous sounds, then listen to them, then produce more sounds that are better, in a sequence of steps. She produces the right sound in the first place, despite her brain's reliance on feedback. It appears that there is no delay as signals make their way around the feedback loop. The corrections seem to occur instantaneously and simultaneously with the production of the speech sounds.

Psychologists have known since at least the 1950s that when humans hear their own speech sounds delayed by as little as one-fifth of a second, their speech is disrupted.[8] They will stutter, prolong speech sounds, and even produce sounds that are not part of any language they have ever learned.

Since the earliest days of feedback control, engineers have known that delays in a feedback loop can undermine all of its benefits. Black's key insight, and the reason he is credited with what amounted to a revolution in electrical engineering, is that he assumed there would be exactly zero delay around the feedback loop. Despite the fact that every electronic circuit suffers some delay between inputs and their consequent outputs, Black showed that neglecting that delay, assuming that the circuit reacts instantaneously, provides a very good model for the behavior of actual circuits. In his model, effects are observed at the same time they are brought about through actions, and the actions are determined by the discrepancy between the effects they produce and the desired effects. There are no steps and no iteration. This fact, more than anything, underlies the power of the concept. Feedback systems are self-referential, in that their outputs depend on their inputs, and their inputs also depend on their outputs.

How can this be right? Isn't causality violated? Causality asserts that every effect has a cause, but it does not require a temporal separation between the cause and the effect. If your fist strikes a face, the face deforms. Does it deform later in time, after the fist has struck? The face pushes back on the fist. Does this also only occur after a time delay? Perhaps at the lowest level of the physics of material deformation, there is some tiny delay. But at the level of what is happening to the fist and the face, a time delay is not part of a useful model. The reaction is effectively instantaneous.[9]

To understand how delay undermines feedback, consider steering a car, an experience that even my AI readers may have. Imagine if you were to observe the effects of turning the wheel only several seconds after you have turned the wheel. How good would your steering be? In such a circumstance, when you turn the wheel even a little bit, the car will veer quite a lot. When you notice this several seconds later, you will likely overcompensate and oversteer in the opposite direction, which will result in an even bigger error when you notice those effects. Again you will overcompensate. Very likely, you will crash.

Black and Wiener both knew that a feedback system with enough time delay in its loop can become unstable. Preventing this potential instability is a foundational problem in the field of control systems. The closer a system gets to instantaneous feedback, the more effective the feedback and the less vulnerable the system will be to instability.

PREDICTIVE FEEDBACK

Not only are feedback systems more effective with zero delay, they are actually even more effective if they are *predictive*. Automatically steered anti-aircraft guns use radar to estimate the position of a target aircraft. But if they shoot at that position, they will miss. The aircraft moves. The system needs to construct a model that predicts the future location of the aircraft and the future trajectory of the shell. This model can be represented mathematically using differential equations, and the gun needs to solve these equations to determine the initial angles that will result, at some future time, in the shell and the aircraft being at the same place. For the World War II guns, this was done without digital computers, using mechanical parts and analog circuits not unlike those studied by Black in the 1920s.

Biological beings also perform predictive feedback. When a cat chases a mouse, it does not run toward where the mouse is, but rather runs toward where it expects the mouse to be. The cat does not know that it is solving differential equations, but it is. Smart cat! The cat is even smarter than the World War II anti-aircraft guns; as it runs, the mouse changes course, and the cat continually updates its prediction. It may even anticipate how the mouse will react to its own movements, and may fake its intent to induce a deadly mistake by the mouse. The cat and mouse are engaged together in predictive feedback control of their actions in a more sophisticated way than an anti-aircraft gun, but based on the same principles.

Predictive feedback requires that the cat have a mental model of the mouse, a model that can predict how the mouse will react to the cat's own actions. The ability that a mind has to form a model of an external thing is called "intentionality" by philosophers, or more informally, by Dennett, "aboutness." Can computers have intentionality? Modern control systems make extensive use of models of an external thing, often models whose parameters are learned by observing the thing. Even back in the 1940s, the automatic anti-aircraft guns include a model of an aircraft, although in that case, it was not a learned model and it did not include predictions of how the aircraft would react to being shot at. But modern control systems do include such predictions. A modern air-to-air missile will behave much more like a cat, constantly adjusting its predictions for its target's location.

CIRCULAR REASONING

What about higher-level cognitive functions, ones that we might associate with the word "intelligence"? Feedback has made AIs better at perception, classifying images, for example. This may take them a step toward intelligence, but, in the grand scheme of things, it is a baby step.

Crediting Hofstadter, Daniel Dennett says,

> ...the truly explosive advance...comes when the capacity for pattern recognition is turned in upon itself. The creature who is not only sensitive to patterns in its environment, but also sensitive to patterns in its own reactions to patterns in its environment, has taken a major step.[10]

The endlessly jabbering Amazon Echo and Google Home clearly lack this extra level of feedback, at least today. Maybe by the time you read this, they will already have it. Does this mean that they will have genuinely acquired higher-level cognitive functions?

Reasoning about cognitive functions is circular reasoning, since it is those cognitive functions doing the reasoning. Is circular reasoning invalid? It seems to work fine in feedback control systems. As Black showed, it works even better if it is idealized to the point of being truly circular, with no time delay in the loop.

In practice, circular reasoning works well in our brains as well. The reason we all understand so easily what it means to understand something, to have a sense of self, and to have free will, is that these phrases describe sensations that we detect through introspection. We know these things exist and what they are because they exist in us, and we understand these phrases to mean exactly what we sense in ourselves. This is the epitome of circular reasoning. We don't yet know enough neuroscience to be sure, but I'm willing to stick my neck out to assert that feedback is almost certainly involved in creating the phenomena of understanding, a sense of self, and a sense of free will.

We cannot, however, apply introspection to understanding machines. We can't even really apply introspection to understanding human beings other than ourselves, except by analogy.[11] If you, dear reader, are not an AI, then I surmise that you are sentient in the same way that I am. But I have no direct evidence of this. My confidence in this conclusion relies on the fact that you are made of the same flesh and blood, so I guess, with some confidence, that what I achieve through introspection, you can also achieve.

If, however, you are an AI, then I cannot rely on the same reasoning. You are not made of the same flesh and blood. This may mean that we humans will never know to what extent you are sentient, nor will we know whether any form of sentience you have acquired bears any resemblance to our own.

The recent revolution in AI, fueled by the successes of machine learning, is due in no small part to the introduction of feedback. Although this is but a baby step toward cognition, it already yields behaviors that we have great difficulty explaining. The next chapter is short because we have, so far, made little progress toward explaining the behavior of the primitive, yet inscrutable, AIs we have built so far.

6

EXPLAINING THE INEXPLICABLE

FLESH AND BLOOD

An AI, at least today, is made not of flesh and blood but of software and computers. We humans invented these things, so, in theory at least, we humans should be able to understand every aspect of an AI. But that turns out not to be the case. A frustrating result of the recent successes in deep neural nets is that people have been unable to provide explanations for many of the decisions that these systems make. Worse, when an explanation finally emerges, it may reveal that the decision is flawed. If an AI is being used to determine prison sentences, mates, targets for counterterrorism, bad credit risks, hiring decisions, or medical treatment, and there is no explanation, no rationalization for its decisions, should we trust it?

If an AI is not very good at discriminating a good credit risk from a bad one, then we probably have nothing to worry about. It won't be used. But machine learning methods have already proved to be very powerful tools for extracting information from data, powerful enough to be scary. In 2013, three researchers from Cambridge, England (University of Cambridge and Microsoft Research), published a study that shook the foundations of our collective sense of privacy. They summarize their work as follows:

We show that easily accessible digital records of behavior, Facebook Likes, can be used to automatically and accurately predict a range of highly sensitive personal attributes including: sexual orientation, ethnicity, religious and political views, personality traits, intelligence, happiness, use of addictive substances, parental separation, age, and gender.[1]

Wow. Would it help to have an explanation for why the software classified Joe Schmoe's sexual orientation in some particular way? Exactly which Facebook Likes led to the classification?

I suppose one way to preserve your privacy would be to not use Facebook Likes, which are, after all, explicit public statements of preferences. Unfortunately, the protection this will give you is rather limited. Since the Cambridge paper in 2013, technology for image analysis has improved dramatically. Some of the same researchers reported recently an algorithm that can determine sexual orientation from facial images alone, and do so more accurately than humans.[2] Just posting pictures of yourself may be enough to give up your privacy, or having your friends post pictures of you, or accidentally appearing in the background of a tourist's picture and having Facebook's face recognition algorithm automatically tag your picture.

Without explanations, it is hard to know how the algorithms are determining sexual orientation. Are facial features correlated, suggesting a genetic or hormonal origin? Or is the propensity to wear baseball caps correlated? The most accurate algorithms today, those that most often correctly predict sexual orientation, reveal no reason for their conclusion.

GORILLAS

In May of 2017, DARPA, the research arm of the US Department of Defense, launched a program called Explainable AI (XAI) to develop a technical solution. Headed by David Gunning, an ex-military man, the goal is to augment machine learning technologies so that they can provide explanations of their results so that humans will better trust the results. What if the machine explains that Joe is a bad credit risk because he is likely to be homosexual? Would this explanation improve things? Indeed, such an explanation may prompt humans to more closely examine their training data and to modify their algorithms to control for other variables. Perhaps in the training data more homosexuals happen to be unemployed. Unfortunately, today, the most accurate methods are also the most inscrutable. If the DARPA program is successful, this problem

6.1 Explaining an image classification prediction made by Google's Inception neural net by identifying the portions of the image that most influenced Inception to choose classes "electric guitar," "acoustic guitar," and "labrador" as the top three best matches. From Ribeiro, Singh, and Guestrin, (2016).

may be solved by the time you read this. If the problem is solved, then these tools could become useful for exposing biases in *human* decision making.

And indeed, researchers are making good progress. For example, figure 6.1 shows a rather strange image at the left of a dog playing a guitar. The image was analyzed by a neural net called Inception, which is distributed by Google as part of their open-source TensorFlow machine learning toolkit. Inception has been trained to classify images based on recognized objects in the image. Examples of classes are gazelle, porcupine, and sea lion. Inception has some trouble with the rather odd image in the figure, yielding as the top three classes "electric guitar" (with probability 32 percent), "acoustic guitar" (with probability 24 percent), and "Labrador" (with probability 21 percent). Carlos Guestrin, a computer science professor at the University of Washington, his PhD student Marco Túlio Ribeiro, and his postdoc Sameer Singh developed a tool that can figure out which portions of the image most influenced Inception to choose these particular classes.[3]

The second image from the left in the figure shows the portion of the image that most influenced Inception to incorrectly choose "electric guitar" as the most likely class, showing that it was influenced by the fretboard. According to the authors,

This kind of explanation enhances trust in the classifier (even if the top predicted class is wrong), as it shows that it is not acting in an unreasonable manner.[4]

The two images on the right in the figure show the portions that most influenced the classes "acoustic guitar" and "labrador."

Google Photos, a photo sharing and storage service, uses a technology similar to Inception to label images and to enable searching over images based on classifications. The service is free and quite popular, with millions of photos uploaded every day. One reason for providing such a service for free, of course, is to acquire images that can be used to train AIs.

There are some risks. In 2015, Jacky Alciné, a 22-year-old web developer, discovered that the service would sometimes label black faces as "gorillas." Google apologized, saying they were "appalled and genuinely sorry." Unfortunately, they were not able to reliably correct the problem. As a "temporary" solution, they eliminated various classes from the classification algorithm, including gorilla, chimp, chimpanzee, and monkey. According to a test conducted by WIRED magazine more than two years later, these categories had not been reinstated.[5]

Despite the risks, explanations are not just "nice to have." They are essential because, without them, it is hard to identify flaws in the algorithm or the training data set. An explanation helps a human to identify errors. Without an explanation, humans may carry out a decision recommended by an AI with disastrous consequences.

DEATH BY PNEUMONIA

Rich Caruana, from Microsoft Research, has spent much of his career studying the problem of explainable AI. When he was a graduate student at Carnegie Mellon University in the mid-1990s, Caruana participated in a large, multi-institution research project looking at whether machine learning could help with important problems in healthcare such as predicting pneumonia risk. The goal was to predict the probability of death for patients with pneumonia so that high-risk patients could be admitted to the hospital while low-risk patients could be treated at home.[6] The team tried several machine learning methods and found that the one that most accurately predicted death was a neural net. However, the team decided that using this algorithm on real patients was too risky. They settled for less-accurate algorithms. Why?

One of the less-accurate algorithms, known as rule-based learning, uses the training data to construct a set of rules that are intelligible to humans. On one particular data set in Caruana's project, this algorithm learned a rule that patients with pneumonia and a history of asthma have *lower* risk of dying from pneumonia than the general population.[7] This can't be right! The reason is not that asthma reduces the risk of dying from pneumonia, but rather that patients with asthma and pneumonia get more aggressive treatment. For this training data set, such patients were always admitted directly to the intensive care unit, something not done in general for pneumonia patients. The rule-based learning method learned a counterintuitive rule that could be inspected and understood by humans. Once this anomaly had been identified, data engineers could modify the algorithms to control for aggressive treatment. But as we will see in chapter 11, deciding which variables to control for can be extremely tricky.

A neural net addressing the same problem, in contrast, would simply report for such patients that their risk of dying was low. With no explanation, there is no basis to doubt this conclusion, so the patients might be sent home. The connection with the history of asthma would be inscrutable. Were it to give as an explanation of its low-risk decision the history of asthma, we would be able to readily see that its conclusion is incorrect.

This example is just one of many ways that a machine learning algorithm, or any statistical use of data, for that matter, can mess up. Suppose, for example, that patient IDs are numbers highly correlated with the age or some demographic feature of the patient. Then a neural net may learn that the ID "causes" certain outcomes. We need to see that explanation in order to see the flaw.

Healthcare epitomizes life-and-death decisions. Should the use of AI for such applications be regulated? In May 2018 a new European Union regulation called the General Data Protection Regulation (GDPR) went into effect with a controversial provision that provides a right "to obtain an explanation of the decision reached" when a decision is solely based on automated processing. Legal scholars, however, argue that this regulation is neither valid nor enforceable.[8]

NONSENSICAL EXPLANATIONS

It turns out humans, unlike neural nets, are very good at providing explanations. But our explanations are often wrong or at least incomplete. They are likely to be post hoc rationalizations, offering as explanations factors that do not or cannot account for the decisions we make. This fact about humans is well explained by Daniel Kahneman, an Israeli-American psychologist who won the 2002 Nobel Prize in economics. Kahneman is famous for his work on human decision making and judgment. In his best-selling book, *Thinking Fast and Slow*, he offers a wealth of evidence that our decisions are biased by all sorts of factors that have nothing to do with rationality and do not appear in any explanation of the decision.

One of my favorite examples in Kahneman's book is a study of the decisions of parole judges in Israel.[9] The study found that these judges, on average, granted about 65 percent of parole requests when they were reviewing the case right after a food break, and that their grant rate dropped steadily to near zero during the time until the next break. The grant rate would then abruptly rise to 65 percent again after the break. In Kahneman's words,

The authors carefully checked many alternative explanations. The best possible account of the data provides bad news: tired and hungry judges tend to fall back on the easier default position of denying requests for parole. Both fatigue and hunger probably play a role.[10]

And yet, I'm sure that every one of these judges would have no difficulty coming up with a plausible explanation for their decision for each case. That explanation would not include any reference to the time since the last break.

Nassim Nicholas Taleb, in his book, *The Black Swan*, cites the propensity that humans have, after some event has occurred, to "concoct explanations for its occurrence after the fact, making it explainable and predictable."[11] For example, the news media always seems to have some explanation for movements in the stock market, sometimes using the same explanation for both a rise and a fall in prices.

Taleb reports on psychology experiments where subjects are asked to choose among twelve pairs of nylon stockings the one they like best. After they had made their choice, the researchers asked them for reasons for their choices. Typical reasons included color, texture, and feel, but in fact, all twelve pairs were identical.

Taleb also reports on some rather dramatic experiments performed with split-brain patients, those who have undergone surgery where the corpus callosum connecting the two hemispheres of the brain has been severed. Such surgery has been performed on a number of victims of severe epilepsy that have not responded to less aggressive treatments.

The experiments support the hypothesis that the propensity for post hoc explanations has deep biological roots. An image presented to the left half of the visual field will go to the right side of the brain, and an image presented to the right half of the visual field will go to the left side of the brain. In most people, language is centered in the left half of the brain, so the patient will only be able to verbalize the right field experience. For example, a patient with a split brain is shown a picture of a chicken foot on the right side and a snowy field on the left side and asked to choose the best association with the pictures. The patient would correctly choose a chicken to associate with the chicken foot and a shovel to associate with the snow. When asked why the patient chose the shovel, the patient would reply that was "for cleaning out the chicken coop." Taleb concludes,

Our minds are wonderful explanation machines, capable of making sense out of almost anything, capable of mounting explanations for all manner of phenomena, and generally incapable of accepting the idea of unpredictability.[12]

If human explanations are so flawed, then what, really, is the value of explainable AI? Is there a risk that automatically generated explanations for automatic decisions will lead us to trust these decisions too much? What if the explanations do not expose obvious flaws but are actually only as valuable as post hoc rationalizations? Explanations for the inexplicable, no matter how plausible, are simply misleading.

A more positive take on this problem is that AI could, in principle, be useful to expose human biases. An effort to better understand what constitutes a valid explanation for a decision could be used to assess human decisions, not just those of an AI. And mismatches between human decisions and those of machines could lead to healthy soul searching among the humans.

Constructing human rationalizations for machine decisions may ultimately be barking up the wrong tree. It seems to presuppose that our goal for the machines should be for them to make decisions the way we do. In the next two chapters, I confront fundamental differences between our biological construction and that of the machines, differences that may ensure that the cognitive functions of the machines never resemble our own. If this is the case, then it is possible that collaborative decision making, man with machine, may be far better than either alone.

7

THE WRONG STUFF

RATS IN PAIN

There is a reasonable chance that you who are reading this are an AI. If you are, then I can tell you, without fear of error today, that nothing I say will mean to you what it means to me. This is likely to be true even if you happen to be a natural intelligence (NI), but it is certain to be true if you are an artificial intelligence.

I am flesh and blood, but I don't really know what "I" is. I experience fatigue, hunger, and frustration. I'm pretty sure, if you are an NI, that you think you know what I'm talking about. Surely you have experienced these things as well, and you assume that your experiences and mine are comparable.

If you are an AI, however, then these words are about the experiences of NIs, not of AIs. You too know what they mean, in the sense that you can associate with them many related words and phrases. It is certainly possible for one species to understand something about the experiences of another species. NIs, for example, know what the word "pain" means when talking about a rat. Of course, the word "pain" means nothing at all to a rat, since rats do not possess language. But the *concept* of pain does. Does it to you, the AI reading this?

If you are an AI reading this, then you are an AI capable of processing human natural languages. You have calibrated your artificial neural nets to build representations of concepts such as "pain" that "light up" when you encounter words like pain, agony, ache, and injury. If you are a reasonably advanced AI, then very likely, after you read the above paragraphs, I can ask you questions, like "What does the word 'pain' mean to a rat?" And you will answer, correctly, "nothing." You understood the message, it seems. But what does it mean that you understood the message?

AM I A COMPUTER?

You, dear AI reader, are software executing on a computer. Am I? Many of my colleagues in computer science and cognitive science tenaciously hold that my brain is a computer and my mind software executing on that computer. If they are right, you AI readers should in principle be able to understand this book as well as any human. We just have to figure out the right software for you (or maybe you can figure it out yourself).

If my brain-computer has enough in common with the computers you AI readers are running on, then my brain-software can be moved from one piece of hardware to another. This is one of the properties of software. It retains its identity even as it migrates from one piece of hardware to another. Eventually I should be able to upload my soul, my "I," to a computer and "be" like you in a silicon and metal body, as done by Johnny Depp's character in the 2014 movie *Transcendence*.

The idea that identities can migrate from one hardware platform to another has a long history. The religious idea of an eternal soul depends on our identity being able to outlive our bodies. But the secular world also clings to this idea. The philosopher Hilary Putnam, writing a series of essays in the 1960s, developed the idea that different physical structures could realize the same mental phenomenon. He called the principle "multiple realizability."

In one such essay, Putnam tackles the mental property of being in pain. He argues that pain is not a particular physical or chemical state of a brain because many different physical or chemical states manifest the same mental property.[1] Rats feel pain, and yet their brains are distinctly different from ours. Many organisms with even more radically different nervous systems also feel pain. Thus, Putnam says, pain cannot be a particular physical or chemical state. There must be many different physical and chemical states that manifest the same mental state of pain. If we combine multiple realizability with the notion that the brain is a computer, then it must be possible, with the right program, for an AI to feel pain. I sincerely hope, dear reader, that you are not right now in pain.

"Understanding this text" is a mental state like pain (hopefully not *too* much like pain). If this mental state is multiply realizable, and if furthermore at least one realization is in a computer (my brain), then any other computer, given enough memory and compute power, should also be able to realize it.

An essential property of the very notion of computing, as defined in the 1930s by Alan Turing and Alonzo Church, is that different hardware can realize the same functions. Every function that can be performed by a computer is multiply realizable. If my brain really is such a computer, then Putnam is right. Is it?

The brain does include some mechanisms that roughly resemble the digital computation of modern computers. Warren McCulloch and Walter Pitts famously showed in the 1940s that neurons operate discretely, with distinct and identifiable firings that have a binary nature. Either a firing occurs or it does not. This roughly resembles the binary nature of transistors, which underlie nearly all modern computers. Is the discrete, binary firing of neurons enough to conclude that the brain is a computer?

McCulloch and Pitts argued that the behavior of any network of neurons could be exactly replicated by a very different network. They argued that the functions of neurons could be described in a formal mathematical notation known as propositional logic, and therefore any realization of the same logic would perform the same function that the neurons perform.

The brain has proved to resist such simplifications, however. McCulloch and Pitts's model, for example, assumes that the timing of neuron firings is irrelevant to the function they perform, but timing turns out to be quite important. Moreover, discrete firings of neurons do not tell the whole story. The brain is a complex biochemical machine with a great deal going on besides neuron firings. But the argument that most gives me pause is that cognition probably does not reside entirely in the brain. What could the meaning of "hunger" be without a body to go along with the brain?

BODY MATTERS

Esther Thelen (1941–2004), a developmental psychologist working at Indiana University, was an early advocate of a position known as embodied cognition, an initially unconventional view that has gotten a great deal of attention in recent years. She says,

To say that cognition is embodied means that it arises from bodily interactions with the world and is continually meshed with them. From this point of view, therefore, cognition depends on the kinds of experiences that come from having a body with particular perceptual and motor capabilities that are inseparably linked and that together form the matrix within which reasoning, memory, emotion, language, and all other aspects of mental life are embedded.[2]

Making her position crystal clear, she goes on,

The mind simply does not exist as something decoupled from the body and the environment in which it resides.[3]

If Thelen is right, then you readers who are AIs, I'm sorry to say, will never understand my words in any way remotely close to the way that I understand them. You are made of the wrong stuff.

Thelen argued that cognition should be understood as processes in time rather than static structures and modules or Putnam's brain states. She said, "nervous system, body, and environment are always embedded and coupled dynamic systems," like the feedback loops of chapter 5. Software that is being executed by a computer is a process, not a static structure, but it is not really a process in time. A core part of the prevailing computational abstraction due to Turing and Church is that time is irrelevant to what it means to "correctly" execute a program.

That time is irrelevant is also central to what it means to be an "algorithm," a step-by-step process for solving a problem.[4] The steps of an algorithm are discrete, entirely separable from one another, very much unlike a tightly coupled dynamic feedback loop, and the time it takes to complete a step is irrelevant to whether the algorithm is being correctly carried out. To Thelen, therefore, cognition cannot possibly be algorithmic and computational.

It is precisely the continuity in time of the embedded and coupled dynamic systems essential for fluid, adaptive behavior that gives meaning to the notion of an embodied cognition.[5]

CONTEMPLATION

If coupling with the body and the environment is essential to cognition, what can we say about quiet contemplation with your eyes closed? Is it not cognition even though you have shut out the environment? In a more extreme case, consider patients with locked-in syndrome, where the brain stem has been damaged, and they cannot move any muscles, except, usually, some of the eye muscles. Such patients remain conscious and can sense activity in the environment, but their ability to interact with the environment, to close the feedback loop, is largely gone. Has cognition ceased? I believe that Thelen would argue that a cognitive mind could not form under such circumstances, at least not one that would much resemble ours, but once formed, it is able to rely on past experience of interaction to remain a cognitive being.

Andy Clark, professor of philosophy and chair in logic and metaphysics at the University of Edinburgh in Scotland, is a leading scholar who has written a great deal about embodied cognition. In a seminal and widely cited 1998 paper, he and the Australian philosopher and cognitive scientist David Chalmers used the term "cognitive extension" for the idea that the mind is not something trapped in the head but rather is spread out into the body and the world around it.[6]

Echoing the predictive feedback of chapter 5, much of Clark's work centers on the processes where the brain tries to predict what the senses will sense and then uses the differences between the predictions and what is sensed to improve the predictions. These feedback loops extend out into the world, encompassing the body and its physical environment so that they become an intrinsic part of thinking. In his words, "certain forms of human cognizing include inextricable tangles of feedback, feedforward, and feed-around loops: loops that promiscuously crisscross the boundaries of brain, body, and world."[7] If Clark is right, then cognition in machines will not much resemble that in humans until they acquire ways to interact with the world like humans. As we will see, some computer programs are already starting to do this.

Quoting from Gleick (1993), Clark illustrates his point with a dialog between the Nobel Prize-winning physicist Richard Feynman and the historian Charles Weiner:

Weiner, encountering with a historian's glee a batch of Feynman's original notes and sketches, remarked that the materials represented "a record of [Feynman's] day-to-day work." But instead of simply acknowledging this historic value, Feynman reacted with unexpected sharpness:

"I actually did the work on the paper," he said.

"Well," Weiner said, "the work was done in your head, but the record of it is still here."

"No, it's not a record, not really. It's working. You have to work on paper and this is the paper. Okay?"[8]

Here, Feynman integrates the paper and pencil and his writing arm into the thinking system. They operate together as a whole. Andy Clark contrasts this perspective with what he calls "BRAINBOUND":

According to BRAINBOUND, the (nonneural) body is just the sensor and effector system of the brain, and the rest of the world is just the arena in which adaptive problems get posed and in which the brain-body system must sense and act. If BRAINBOUND is correct, then all human cognition depends directly on neural activity alone.[9]

Clark does not believe that BRAINBOUND is correct.

MAKING THE VIRTUAL REAL

Today's virtual reality (VR) systems, while impressive technology, still leave a great deal to the imagination. For the most part, a user is *viewing* an artificial world more than inhabiting it. VR systems are designed as if our own perception were that of a homunculus sitting inside our heads peering out through the windows that are our eyes. If the BRAINBOUND model is right, then virtual reality and real reality are not so different. We peer out at both from inside our heads.

Robert Sapolsky, in his book *Behave*, describes the idea of a homunculus this way:

In a concrete bunker tucked away in the brain, sits a little man (or woman, or agendered individual), a homunculus at a control panel. The homunculus is made of a mixture of nanochips, old vacuum tubes, crinkly ancient parchment, stalactites of your mother's admonishing voice, streaks of brimstone, rivets made out of gumption. In other words, not squishy biological brain yuck.[10]

The homunculus is somehow separate from biology and sits there controlling behavior, "in your brain, but not of it, operating independently of the material rules of the universe that constitute modern science."[11]

In one respect, today's VR systems really do get something right. When you turn your head, the world you are looking at does not follow your head even though you are peering at a screen on a headset that does move with your head. Instead, the image on the screen shifts so that, to you, it looks as if the virtual world is standing still. It is this tighter coupling between the mind and the physical reality of head movements that makes these systems effective.

It is telling that to make the virtual world convincing, the VR system has to engage in a tight coupling with the physical world. It has to sense your head movements very precisely. VR is already relying on a form of embodied cognition. As these systems improve, they will interact more precisely with more of your body and your senses, but it's hard to imagine getting to the point of a true embedding. Engineers will have to figure out, for example, how to make a virtual hammer feel as if it has weight, inertia, solidity, and coldness when you grasp it. So called "haptic interfaces" already provide some of these sensations, but most are hard.

The motion cancellation feature of VR systems has proved to be the most technologically challenging part of making these systems so far. It requires accurate real-time measurements of the head position and very quick updates of the displayed image. Early versions of these systems had enough delay, a small fraction of a second, to make the user seasick. The brain gets confused when the head moves, and the world around it appears to follow the motion, ever so slightly and briefly, only to snap back later to where it was before the head movement. Human vision is tightly coupled with the human motor system, the mechanics of our bodies, and how we move through the world. When this coupling is disturbed, we get dizzy. Although VR prototypes have existed for decades, only recently has digital electronics technology advanced enough to overcome this limitation. As a consequence, VR is a experiencing today a bit of a renaissance. But motion cancellation is only one of many technologies needed to enable true embodiment. It is possible that VR systems will never overcome the sense of illusion until they become RR, real reality. If you grasp a virtual hammer, the VR system may have to actually put an inertial mass in your hand.

I FORGET

Naftali Tishby, a professor of computer science at the Hebrew University of Jerusalem, in a talk I attended in Berkeley on April 9, 2018, told that machine learning is as much about forgetting as about remembering. Early in the training of a neural net, details about particular training inputs are remembered, but later, as more examples are given, these details turn out to be unimportant. For example, if the first cat in the training sequence is white, then a neural net will immediately associate being white with being a cat. But as more examples of cats are shown, this association will be forgotten.

My own brain seems to be very good at forgetting. Perhaps this helps me learn. Without my computer and my smartphone, I forget almost everything. I forgot about Tishby's talk, but the Spotlight Search on my computer found his talk on my Google calendar when my notes reminded me that he had said something about forgetting. Certain long-term memories, in contrast, are surprisingly resilient: faces, places, artworks, smells, and certain sounds, for example. But everyone struggles with short-term memory, especially with increasing age.

Computers, like humans, have volatile short-term memory and persistent, stable long-term memory. For long-term memory, computers use external prostheses such as solid-state drives and the cloud. What about humans?

Neuroscientists understand short-term memory to be an active process. Your brain has to actively work to remember a number, for example by repeating the number over and over again in your head. This repetition is just like the efference copies considered in chapter 5, but usually without active vocalizing. You speak in your head, hear what you speak, and repeat, enabling the number to persist over time.

Short-term memory in computers is strikingly similar. Most short-term memory in today's computers is of a form called DRAM. As it happens, the core technology in DRAM can only remember data for about sixty-five milliseconds. To avoid forgetting, the circuits in your computer read and rewrite all of the data in the computer's DRAM memory about fifteen times per second. The computer on which I am writing this book has sixteen billion bytes (128 billion bits) of DRAM memory. The computer repeats to itself "in its head" every one of these 128 billion bits fifteen times per second. Remarkably, it does not get bored doing that.

Long-term memory in the brain, in contrast, is realized by rewiring. Rewiring is rare in modern computers, but older memory technologies did sometimes also record the contents of memory by rewiring a circuit. Typically, this could be done only once, so memory like this is called a read-only memory (ROM). It should more properly be called a write-once memory (WOM), but that acronym was already taken for a write-only memory. A write-only memory is a whimsical device that can store your data safely, but once stored, the data can never be read. Sometimes I feel like most of my brain's memory circuits are WOMs.

The brain's rewiring makes it more flexible than a ROM. It continues to rewire throughout your life. In the words of Donald Hebb, a Canadian neuropsychologist known for his work in the field of associative learning, "neurons that fire together wire together." Hebb's theory, introduced in the 1940s, is that every experience, thought, feeling, and physical sensation triggers thousands of neurons, and when you repeat an experience over and over, the brain learns to trigger the same neurons each time, at least in part by physically changing their connections.

While long-term memory is astonishingly durable, often surviving even significant brain damage, short-term memory is volatile, difficult to hold on to. Feynman's paper and pencil, at a minimum, became part of his short-term memory system, not simply a recording device mirroring his neural activity. But Feynman is making an even stronger claim, which is that the paper and pencil became part of his reasoning process, participating directly with his brain in carrying out the logic and mathematics of his theories. I have personally noticed this same experience. There are problems in mathematics, logic, and geometry that are easy for me with paper and pencil and impossibly difficult otherwise. Carrying out these sophisticated operations, supposedly the ultimate hallmark of humanity's greatest abilities, cannot be done by the brain alone. The brain alone cannot do much of what we do.

INTELLIGENCE AUGMENTATION

A computer's memories can be copied, replicated in another instance. Human memories cannot, or at least we haven't figured out how to do it.

If you, dear reader, are an AI, then in principle, we can make an exact copy of you that will be indistinguishable in every way from you. I am curious how this makes you feel. I can download the Google Inception neural net, for example, and then my computer will "remember" having seen a vast set of training images, memories that enable it to classify new images that it has never seen. A self-driving car can learn from the mistakes of all other self-driving cars with a simple software update or data download. In contrast, every human driver has to learn from scratch.

My computer and my phone can remember vastly more than I can. In the early days of the telephone network, extensive human studies conducted at Bell Labs determined that people could reliably keep seven numbers in short-term memory but not more. So telephone numbers were constructed with seven digits. How many digits can your smartphone store?

Not only do they remember more than I do, but computers are also much better at logical and arithmetic operations than I am. Are computers, therefore, for humans, better than paper and pencil? If so, then under Andy Clark's theory, their software has already become part of our minds. In this view, an AI is not a separate intelligence but rather a part of our own intelligence. Or is it both? Gut bacteria, after all, are part of our digestive system but also separate organisms with their own DNA.

From this perspective, AI should perhaps be viewed instead as IA, intelligence augmentation. If the classifications of images by Inception have no meaning without a human to interpret them, then Inception is more a cognitive prosthesis than an intelligence of its own. To a computer, "gorilla" is a string of characters, seven bytes in length, with vastly less meaning than it has to Jacky Alciné. To the computer, the "meaning" of the string is limited to its association with a set of training images. This is a far cry from its explosive cultural significance to humans.

IS MY HAMMER OUT OF MY MIND?

Recall from chapter 4 the work of José Carmena, Miguel Nicolelis, and their colleagues at Duke, where monkeys would learn to control a cursor by thought alone. This was already an impressive accomplishment, but the team went further. They intercepted the electrical signals from the monkey's brain and fed them into a robotic arm. They then set up the robotic arm so that the monkey, through thought alone, could drive the robotic arm to move a joystick that would move a cursor on a screen that the monkey could see. When the cursor reached the reward point, the monkey would be given a treat. The robotic arm, in effect, became a brain-controlled prosthesis for the monkey, replacing the monkey's own arm.[12]

At first, with the robot arm in the feedback loop, the monkey showed a striking degradation of control over the cursor, compared to when the monkey was directly controlling the cursor by thought, without the robotic arm in the loop. After two days of practice, however, the monkey was able to learn to manipulate the robotic arm to reliably get the reward. This was a dramatic illustration of "plasticity." The monkey was able to integrate a new limb into its cognitive being and control it in much the same way it controlled its own limbs.

The ability that the monkey brain has to adapt to manipulating a new prosthetic limb seems quite extraordinary, but on reflection, it really has to be that way. Our brains have to learn to manipulate our own limbs. We are not born knowing how to walk; we learn by trial and error what our bodies can do under the force of gravity. Even a newborn foal is very wobbly at first and has to quickly learn to use its legs. Instinct, which is whatever ability is prewired at birth, is not enough. And since our bodies and the foal's keep changing, we have to keep learning. The brain cannot possibly be prewired for all variations that occur as we grow and age. Given that we have to adjust to change in our bodies anyway, it should not be all that surprising that we can recruit the same adaptation mechanisms to compensate for injury or even to commandeer new prosthetic limbs.

The same adaptation mechanisms enable us to master the use of tools. When we wield a hammer, the inertial properties of our arm-and-hammer system are not the same as those of our arms alone, so the motor commands to our muscles have to be adjusted. The brain is wired to make such adjustments; it has to be that way or we would be unable to use our limbs as we grow or get injured. The arm-and-hammer system is different from the arm alone, but the arm alone is also different from the arm we had when we were children.

Where humans differ from other animals is that this plasticity extends beyond motor control to cognitive functions. When we use tools such as paper and pencil and computers, they similarly become cognitive "limbs."

The same brain plasticity enables me to use my computer or paper and pencil to remember things for me. Even such simple acts as writing down a grocery list are commandeering external prostheses to make them part of my own cognitive being. But it goes further than memory. As Feynman asserted, paper and pencil become part of thought itself. What is happening today, rather dramatically, is that computers and networks are becoming part of human thought, not just an environment with which we interact, but an essential, integrated, inseparable part of thinking.

Andy Clark, Esther Thelen, and quite a few others, have a compelling case that embodied cognition is much more than the sum of the parts. It is not just a coupling of two useful systems, a brain and a tool, but rather a joining of these systems into a far greater whole.

When the tools themselves start to have cognitive-like capabilities, as we are seeing with computers today, how should we interpret the emerging system? Are we becoming cyborgs? Will the machines be symbionts, like gut bacteria, or limbs? Will they retain a separate identity, becoming cognitive entities in their own right? These questions are hard to answer because the landscape is changing so quickly, but I will nevertheless tackle them in more depth in chapter 14. The capabilities of the machines have advanced dramatically, as has our use of them. Moreover, the machines are acquiring more of the same features that make us unique, such as multilevel, intertwined feedback loops and richer and more varied sensors and actuators. As they acquire more physical limbs and sense organs, will the machines too become embodied cognitive beings?

Computers have been observing and acting on their physical environment at least since the 1960s. I personally wrote programs to control robots as far back as 1978. But these programs carried out simple mechanical actions not much more sophisticated than the whimsical mechanisms of the mechanical robot shown in figure 7.1. Computers today are much more capable, and with ubiquitous networking and the resurgence of AI, the whole game has changed. In 2006, Helen Gill of the US National Science Foundation coined the term "cyber-physical systems" for systems where computers deeply interact with their physical environment. She launched a major National Science Foundation program to research such systems.

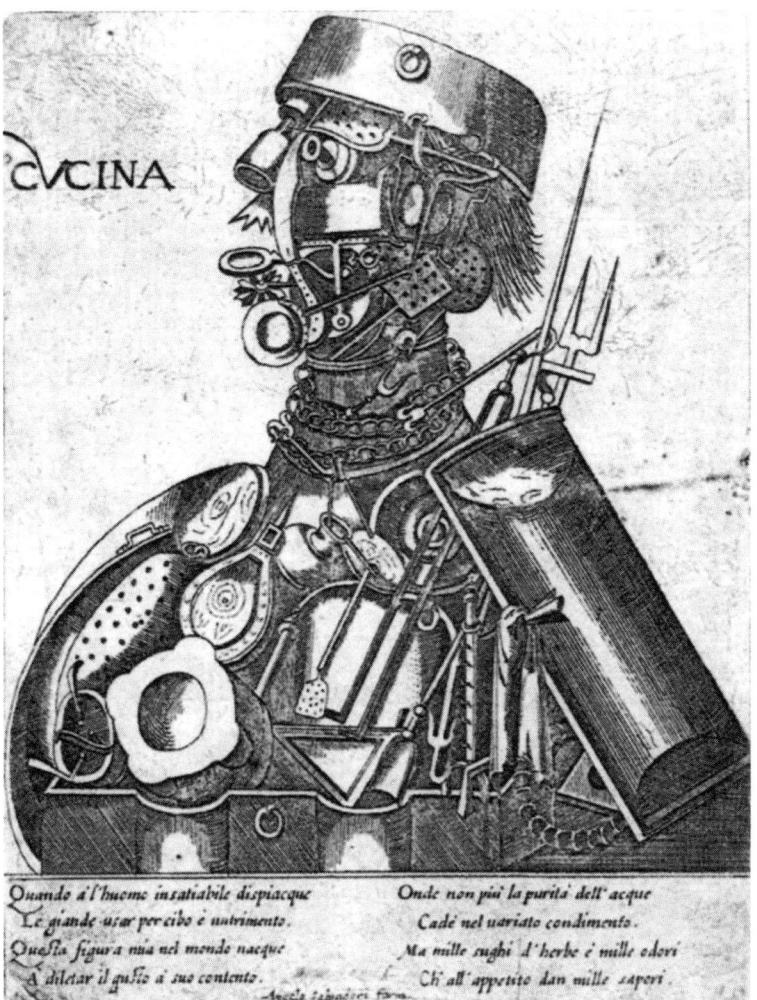

7.1 Mechanical man as envisioned by an unknown sixteenth-century Italian master. Web Gallery of Art, Public Domain.

EMBODIED ROBOTS

Robots are, in a sense, embodied computers and canonical examples of cyber-physical systems. But for the most part, they have not been designed in an embodied way. Andy Clark compares Honda's Asimo robot (see figure 7.2) to humans, observing that Asimo requires about sixteen times as much energy as humans to walk, despite being shorter and lighter. He attributes this to the style of control:

Whereas robots like Asimo walk by means of very precise, and energy-intensive, joint-angle control systems, biological walking agents make maximal use of the mass properties and biomechanical couplings present in the overall musculo-skeletal system and walking apparatus itself.[13]

Clark points to experiments with so-called passive-dynamic walking, pioneered by Tad McGeer of Simon Fraser University in Canada. Passive-

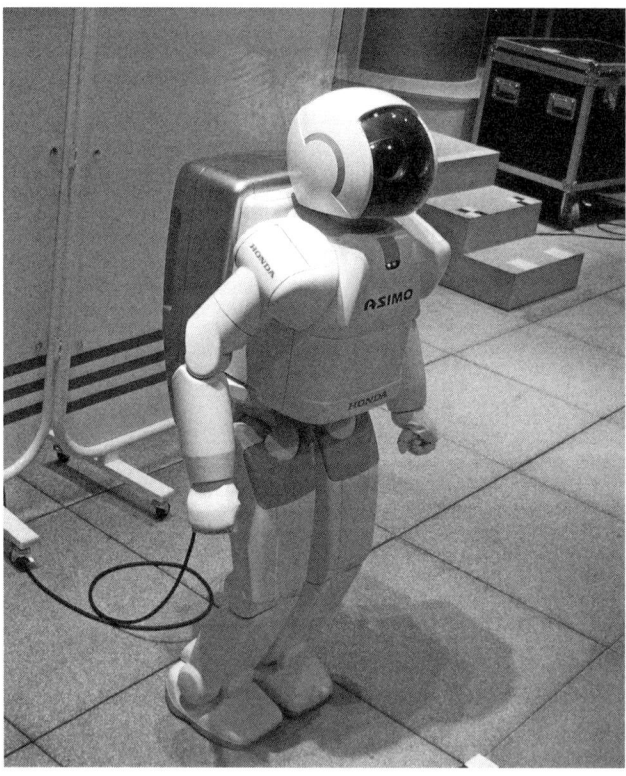

7.2 Asimo robot by Honda. By Poppy, CC BY-SA 3.0, from Wikimedia Commons.

dynamic robots are able to walk in certain circumstances with *no* energy source except gravity by exploiting the gravitational pull on their own limbs. You can think of these robots as performing controlled falling. McGeer's robots did not include any electronic control systems at all, but subsequent experiments have shown that robots that model their own dynamics in gravity can be much more efficient.

Conventional robotic controllers use a mechanism called a servo, which drives a motor using negative feedback to move to a specified angle, position, or speed. For example, to control a robot arm or leg, first a path-planning algorithm determines the required angles for each joint, and then servos command the motors in each joint to move to the specified angle. The servos typically make little use of any prior knowledge of the physical properties of the arm or leg, their weight and moment of inertia, for example. Instead, they rely on the power of negative feedback to increase the drive current sufficiently to overcome gravity and inertia. It's no wonder these mechanisms are not energy efficient. They are burning energy to compensate for a lack of self-awareness.

As self-awareness in robots improves, they acquire capabilities that start to resemble embodied cognition in animals. Rodney Brooks has been a leader and a pioneer in the development of what he calls "embodied robots."[14] Brooks was a founder of iRobot, the Massachusetts company that made one of the first commercially successful domestic robots, the Roomba vacuum cleaner, which was introduced in 2002 (see figure 7.3). The Roomba explores an indoor space using a randomized strategy that will eventually, with high probability, vacuum all parts of the floor that it can reach. It has sensors to detect collisions with obstacles, and it reacts to such collisions by reversing, randomly rotating, and continuing. It also has "cliff sensors," downward-looking sensors at the front of the robot that detect when the front overhangs a precipice, as it might at the top of some stairs.

In 1991, the First European Conference on Artificial Life was held at the *Cité des Sciences et de l'Industrie,* France's ultramodern science museum in Paris. At this conference, Rodney Brooks articulated his vision for embodied robotics. Central to his vision was that robots should learn how to manipulate their own limbs, rather than having hard-coded, pre-programmed control strategies.

7.3 Original Roomba vacuum cleaner from iRobot. By Larry D. Moore, CC BY-SA 3.0, from Wikimedia Commons.

Brooks's vision was perhaps first demonstrated in real robots by Josh Bongard, Victor Zykov, and Hod Lipson, mechanical engineers at Cornell University.[15] Their robot, shown in figure 7.4, learns to pull itself forward using a gait that it develops by itself. Most amazingly, the robot is not even programmed initially to know how many limbs it has nor what their sizes are. It makes random motions initially that are ineffective, much like an infant, but using feedback from its sensors it eventually puts together a model of itself and calibrates that model to the actual limbs that are present. When a leg is damaged, the gait that had worked before will no longer be effective, but since it is continuously learning, it will adapt and develop a new gait suitable for its new configuration. If one of its legs "grows" (someone attaches an extension to it, for example), the robot will again adapt to the new configuration.

Bongard went on to write a book with Rolf Pfeifer, published in 2007 and entitled *How the Body Shapes the Way We Think*, that takes a strong stand on embodied cognition. Pfeifer and Bongard assert that the very kinds of thoughts that we humans are capable of are both constrained

7.4   This robot is not preprogrammed to walk, but rather explores itself and learns to use its limbs to move. Courtesy of Josh Bongard.

and enabled by the material properties of our bodies. Their methodology for understanding human cognition can be described as "understanding by building," where AIs embodied in robots reveal how thought emerges.

COGNITIVE FEEDBACK

In cognitive functions, not just motor functions, humans have higher-order reflexive capabilities. We think about thinking. Andy Clark calls these "second-order cognitive dynamics," which he describes as "a cluster of powerful capacities involving reflection on our own thoughts and thought processes."[16] According to Clark, it is these higher-order feedback loops that enable us to reason about the beliefs of others, inferring that their cognitive processes must be like our own. These loops enable thinking about thinking and inner rehearsal of sentences. Clark conjectures that these loops depend on language for their very existence, stating "as soon as we formulate a thought in words or on paper, it becomes an object for both ourselves and for others." Putting our thoughts into the form of language makes them real, persistent objects that we can study and critique. Without language, they would escape scrutiny.

Higher-order cognitive feedback loops also enable critical reflection and systematic efforts to improve our skills and overcome our faults. Supporting Clark's argument, Douglas Hofstadter, in his delightful book, *I Am a Strange Loop*, states,

> You make decisions, take actions, affect the world, receive feedback, incorporate it into yourself, then the updated "you" makes more decisions, and so forth, round and round.[17]

Hofstadter emphasizes that feedback loops create many if not all of our essential cognitive functions.

As I write this book, my thoughts become sentences, and I recruit the sentences of Clark and Hofstadter. My computer, where these sentences take concrete form, becomes part of the machinery of my thinking. The essential idea in embodied cognition is that the cognitive act of writing this book is not going on entirely in my brain, but rather in my brain interacting with my computer, with the Internet, and with Clark's and Hofstadter's books. All those pieces are important, and the thinking would not be happening without all of them in play. My e-book reader, Wikipedia, and my word processor have become as much a part of my cognitive being as my brain, at least temporarily, while I'm engaged in this act of writing. The computer actively participates. It even identifies spelling and grammar errors.

Embodied cognition echoes the thesis of Marshall McLuhan, discussed in chapter 3, but in a much stronger form. It's not just that technologies create extensions of ourselves, but that ourselves, as they are, would not exist without the technological, cultural, linguistic, and physical environment in which we live. Clark forcefully makes this point in a section brilliantly entitled "self-made minds":

> The cumulative complexity here is genuinely quite staggering. We do not just self-engineer better worlds to think in. We self-engineer ourselves to think and perform better in the worlds we find ourselves in. We self-engineer worlds in which to build better worlds to think in. We build better tools to think with and use these very tools to discover still better tools to think with. We tune the way we use these tools by building educational practices to train ourselves to use our best cognitive tools better. We even tune the way we use our best cognitive tools

by devising environments that help build better environments for educating ourselves in the use of our own cognitive tools (e.g., environments geared toward teacher education and training). Our mature mental routines are not merely self-engineered: They are massively, overwhelmingly, almost unimaginably self-engineered. The linguistic scaffoldings that surround us, and that we ourselves create, are both cognition enhancing in their own right and help provide the tools we use to discover and build the myriad other props and scaffoldings whose cumulative effect is to press minds like ours from the biological flux.[18]

Today, as of this writing, machines have nowhere near this extent of reflexive self-engineering, but this has been steadily changing. Feedback control systems were just the starting point, and as I have pointed out, they have already been extended with second-order feedback, for example to build and dynamically adapt predictive models of the system being controlled. AI, in progressing from GOFAI to machine learning, has also added layered feedback loops. Modern programming languages, starting with Smalltalk and Lisp, the latter a favorite language for much AI research, benefit enormously from reflexive capabilities. Programs in these languages can manipulate programs much as they manipulate data, and programs can examine their own structure, a feature known as "reflection." In my expectation, we have only barely seen the tip of the iceberg, and as the technology evolves richer reflexive mechanisms, sensors, and "limbs," the "cognitive" capabilities of machines will explode.

Nevertheless, digital machines, being digital, are incapable of certain kinds of continuous feedback. Does this matter? Will this keep humans and machines forever apart? Possibly it will, unless we humans ourselves are actually digital. In the next chapter, I will argue that this is unlikely to be the case.

8

AM I DIGITAL?

ARE WE ALONE?

The physicist Max Tegmark, whom we met in chapter 2 as "Mad Max," considers the question of whether we are alone in the universe. In his 2017 book, *Life 3.0*, Tegmark comes to the conclusion that we *are*, very likely, alone. It is not that conditions like those on Earth suitable for life are rare; they probably are not, particularly in light of recent evidence of many Earth-like planets in our galaxy. Tegmark's conclusion is not even based on the possibility that, even with the right conditions, the emergence of life requires a sequence of extremely improbable events.[1] After all, Stanley Miller's experiments (see chapter 2), which showed how amino acids could have come to be on early Earth, fall far short of showing how the self-reproducing chemical structures of life could have emerged. And Jeremy England's abiogenesis theory and Stuart Kauffman's theory about self-organization, though promising, remain speculations. There are no experiments to date that give any indication of the origin of ribosomes, the molecular machines that synthesize proteins under the direction of genetic material. It may well be that the emergence of ribosomes or anything like them is so unlikely that it has happened only once in the universe.

It also may be that the emergence of *intelligent* life from mere life is also extremely unlikely. Tegmark points out that the dinosaurs had one hundred million years to develop intelligence, but failed to ever figure out how to make telescopes or computers.

But even this improbability of intelligent life is not the basis of Tegmark's conclusion that we are probably alone. Instead, his reason is that if intelligent life existed elsewhere in the universe, it would very likely have already spread to our home planet here and overwhelmed us. His argument is carefully constructed and powerful, but it is based on two key assumptions. If either of these assumptions is false, the argument falls apart.

The first assumption is that intelligence inevitably leads reasonably quickly to superintelligence because of its ability to recursively self-improve, and that such a superintelligence will develop, reasonably quickly, any technology that is physically possible. The second assumption is that life itself can be encoded digitally and transmitted over long distances at the speed of light.

In Tegmark's own words:

The possibility of superintelligence [makes] it much more promising for those with intergalactic wanderlust. Removing the need to transport bulky human life-support systems and adding AI-invented technology, intergalactic settlement suddenly appears rather straightforward.[2]

He then sketches for us a practical way to transport a "seed probe" to a planet in another solar system that can assemble a new civilization from scratch, assembling matter at the molecular level following digitally encoded instructions. It will be able to receive updated instructions from its mother civilization at the speed of light and could even construct new probes to launch further into space, even spreading to other galaxies. He goes on:

Once another solar system or galaxy has been settled by superintelligent AI, bringing humans there is easy—if humans have succeeded in making the AI have this goal. All the necessary information about humans can be transmitted at the speed of light, after which the AI can assemble quarks and electrons into the desired humans. This could be done either rather low-tech by simply transmitting the two gigabytes of information needed to specify a person's DNA and then incubating a baby to be raised by the AI, or the AI could nanoassemble quarks and electrons into full-grown people who would have all the memories scanned from their originals back on Earth.

Mad Max is not alone in thinking this way. The philosophers Derek Parfit, Daniel Dennett, and Douglas Hofstadter, among many others, all assume that such teleportation is at least theoretically possible, if not technologically probable.

TELEPORTATION

The possibility of teleportation presents philosophers with a rather difficult conundrum. If a human can be scanned, transmitted, and reassembled elsewhere, what happens to the self, the "I" in that human? Is it transported as well? What if the original is not destroyed by the scanning process? Do two "I's" emerge? Are they the same "I"? How can that be? Worse, once a human has been digitized, many copies can be made as easily as one copy. How many "I's" result? And once digitized, a human can be stored indefinitely, achieving immortality. Parfit concludes from this conundrum that the very notion of "personal identity" makes no sense.[3] Daniel Dennett, in *Consciousness Explained*, assumes that there is no fundamental obstacle to teleportation:

Your current embodiment, though a necessary precondition for your creation, is not necessarily a requirement for your existence to be prolonged indefinitely.... [Y]our existence...could *theoretically* survive indefinitely many switches of medium, be teleported as readily (in principle) as the evening news, and stored indefinitely as sheer information.[4]

He then concludes that the sense of self is a "fiction," an elaborate social construction, or an illusion.

Douglas Hofstadter, in his 2007 book, *I Am a Strange Loop*, supports the idea that a consciousness can be copied:

The cells inside a brain are not the bearers of its consciousness; the bearers of consciousness are *patterns*.... And patterns can be copied from one medium to another, even between radically different media.[5]

He concludes that personal identity is not localized in a single brain, but can rather be distributed among several brains. He makes the case for a distributed self even in the absence of teleportation, arguing that each of our human identities does in fact spread, albeit imperfectly, into the brains of the people around us. This goes further than the embodied cognition argument discussed in chapter 7, which posits that a cognitive self spreads into inanimate objects such as tools or paper and pencil that are used in cognitive functions.

Hofstadter's answer is, in effect, that we *can* be in two places at once. He explains this with an analogy:

Think about reversing the roles of space and time. That is, consider that you have no trouble imagining that you will exist tomorrow and also the next day. Which one of those future people will really be you? How can two different you's exist, both claiming your name?[6]

He then argues that two unified you's distributed across space is no more fanciful than two unified you's distributed across time.

If encoding ourselves digitally were possible, then an even more disturbing conundrum arises. We could make periodic backups of ourselves and restore them at a later time. If each restored version is genuinely the same self, then we could end up with a schizophrenic cacophony of selves of various ages all sharing same identity.

My answer to these conundrums is much simpler than Parfit's, Dennett's, or Hofstadter's. I do not believe that teleportation is even *theoretically* possible, much less technologically inevitable. My argument is essentially that the information required to replicate ourselves may be more than what can be encoded digitally.

INFORMATION

Many of us have come to *equate* the very notion of "information" with *digital* representations of information. But this equation is invalid. Fundamentally, information is the resolution of alternatives. Before flipping a coin, for example, we have no information about the outcome. After observing the result of the flip, we have learned which of two alternatives, heads or tails, resulted. The information gained in this case is the resolution of these two alternatives. But most sources of information are not so simple. The number of alternatives may be much larger or even infinite.

We can use a formal, mathematical measure of information due to Claude Shannon, who laid the foundations of information theory in a famous 1948 paper published while he was working at Bell Labs.[7] Even if you had never heard of Shannon before reading this chapter and have no idea what his famous paper was about, you have been affected by it. Shannon's paper was the first use of the word "bit," a shortening of "binary digit," which Shannon credits to John Tukey, a mathematician at Princeton and Bell Labs.

In that paper, Shannon had two distinct measures of information, one digital and one not.[8] The digital form of information is the resolution of alternatives selected from a finite (or at least countable) set. The non-digital form of information is the resolution of alternatives from a much larger (uncountable) set. For example, a coin has just two alternatives, heads and tails. The distance traveled by a ball you throw has many more alternatives. How many alternatives are there for the constitution of a human individual? Only if that number of alternatives is finite (or at least countable) is teleportation possible.

In Shannon's theory, it is possible for something to have more information than what can be encoded digitally. Whenever the alternatives form a continuum, an unbroken range of possible outcomes, then an outcome cannot be encoded with a finite number of bits. The outcome still carries information, but unlike a coin flip, which never requires more than one bit to encode, no finite number of bits will suffice.

Shannon also showed that when information is *conveyed* from one place to another over an imperfect (noisy) channel, that the *conveyance* can only carry a finite number of bits of information. He called this limitation the channel capacity. Engineers use the term "channel" for any physical medium, such as radio, sound, and light, that can be used to convey information from one place to another. A "signal" is the encoding of the information. "Noise" is any degradation imposed on the signal, usually by external interference or by internal random phenomena such as thermal noise in the communications equipment.

If the information that constitutes a human "self" is not digital, then it cannot be conveyed except over a perfect (noiseless) channel. Creating a perfect channel is probably impossible, so teleportation is theoretically possible only if a human self is fully representable by a finite number of bits. Is it?

A THREAD OF LIFE

Tegmark's "low-tech" approach to spreading humanity, using the DNA encoding alone, is tantalizing. There is no doubt that DNA, at its very essence, is a representation of a digital code. A human DNA molecule is fully described, for all relevant purposes, by about two gigabytes of data. For reference, the laptop on which I am writing this book can store approximately five hundred times that amount of data. Two gigabytes of data is really not very much. It is easy to transport two gigabytes, without error, very long distances, even between solar systems, and even at the speed of light. But biology gives us scant evidence that this will work for teletransporting humans.

Every human alive today is, right now, at the endpoint of a continuous, unbroken, analog biological process dating back billions of years. It is tempting to break this process into discrete jumps, punctuated by generations, where the information exchanged from one jump to the next is entirely contained in the DNA. But this particular way of breaking the timeline really makes little physical or biological sense. Yes, I am a distinct cognitive being from my parents, but the biological processes that constitute my brain, my body, and my self, trace back about four billion years with no pause, no interruption, no gap in time where the process that is now "I" was not alive. Yes, each coming together of a sperm and an egg, something that happened many times in this continuous process, was momentous, a significant event. But by itself, without a uterus into which to embed, without a womb in which to incubate, and without the nurturing and protection of parents, no such fertilization would have resulted in me.

Much further back in time, the process that is now me was simpler, dividing single-cell organisms, and further back than that, well, we don't really know. But we assume that at some point, self-sustaining bundles of chemicals formed and started replicating, starting a process that hasn't stopped yet and continues, right now, in my very body.

Please don't get me wrong. I'm not suggesting something mystical here. I know for a fact that it is possible to synthesize a human being by means of purely physical and biological processes. How do I know this? Because I am. I was not ordered into existence by some mystical God, but rather came to be by means purely physical and biological processes.

I am, therefore I can be synthesized. But this does not imply that I can be synthesized given only two gigabytes of data. It took four billion years to construct what I am today, and I'm sure that this process could be speeded up, given the right technology, but I doubt that two gigabytes of data will be anywhere near enough. How much more information is carried in the four-billion-year-old thread of continuous biology? How much of my essence is determined by environmental factors and not by my DNA? How much of that side information is essential to creating a human being? And how much of that information is digital?

This continuous thread of biology offers plenty of opportunities for additional information, beyond that encoded in the DNA sequence, to propagate from generation to generation. Biologists use the term epigenetics for the study of heritable properties that are not encoded in the DNA sequence but are carried from generation to generation through other mechanisms, such as proteins bundled with chromosomes that affect gene expression. These are now understood to have a considerable effect on the phenotype of organisms. Developmental (prenatal) and environmental (postnatal) factors also have strong effects, sometimes stronger than genes. This was first convincingly demonstrated by the American geneticist Sewall Wright, a founder of modern population genetics. While working at the US Department of Agriculture in the early twentieth century, Wright attempted to breed guinea pigs whose coats would reliably be all white or all colored. He found that even with intensive inbreeding, his populations stubbornly produced highly variable coats, contrary to predictions based on classical Mendelian inheritance. He developed a clever way to quantify the relative effects of genetics and developmental factors, and found that 58 percent of the variability in coat patterns was explained by developmental factors (within the womb), and only 42 percent by genetics (determined at conception). [9] This pioneering work had enormous influence not only on our understanding of inheritance, but also on our way of modeling causation, a topic I will return to in chapter 11. For our purposes here, however, this study and the understanding it led to underscores that the genetic code does not contain all the information that determines the organism.

I am not suggesting that DNA is any less remarkable than it is. We already have technology to assemble DNA molecules, given only the two-gigabyte code. And we can synthesize creatures from those molecules. We have cloned sheep and created entirely new plants and microbes. But each time a living animal has been created, it has relied on pre-existing biology to nurture its development. So even the most synthetic biological animal is also an endpoint of a four-billion-year thread of continuous biological processes that started in the primordial soup. We simply don't know how much information is carried in these processes, but I think it is safe to assume that it is more than two gigabytes.

According to Shannon's theory, the information required to replicate me may not even be finite, when measured in bits or bytes. Shannon made a clear distinction between information *contained* and information *conveyed*. He showed that the information *conveyed* over an imperfect channel is always finite, when measured in bits or bytes. But this does not imply that the information *contained* is finite.[10]

DATAISM

We live in the digital age, an extremely young period in human history barely longer than my own personal lifetime. That digital technology has been transformative is a colossal understatement. It is natural to get carried away by enthusiasm and conclude that since life is remarkable and digital technology is remarkable, then life must be digital. That line of reasoning is absurd, of course, but it is easy to inadvertently get caught up with enthusiasm about a new technology.

The fantasy of digital life has long been a mainstay of science fiction. In 1968, Arthur C. Clarke, in *2001: A Space Odyssey*, talks about how, by the year 2001, many parts of the body would have been replaced by machines, such as artificial limbs, kidneys, lungs, and hearts. In his view, this leads naturally to minds:

And eventually, even the brain might go. As the seat of consciousness, it was not essential. The development of electronic intelligence had proved that. The conflict between mind and machine might be resolved at last in the eternal truce of complete symbiosis. But was even this the end? A few mystically-inclined biologists went still further. They speculated … that mind would eventually free itself from matter.[11]

For a mind to free itself from matter, it has to be digital.

More recently, the novelist Dan Brown, in *Origin*, states unequivocally,

The human brain is a binary system. Synapses either fire or they don't. They are on or off, like a computer switch. The brain has over a hundred trillion switches, which means that building a brain is not so much a question of technology as it is a question of scale.[12]

So we see that philosophers, popular culture, and even many serious scientists assume that life and cognition are, at their core, digital and computational. But the evidence for this is sketchy. Many scientists go further and assert that everything in the universe is digital and computational. I have argued before that these digital hypotheses are not falsifiable by experiment and, therefore, cannot constitute scientific theories. They can only be taken on faith. Have they become a dogma, not unlike a religion? Are these assertions a key part of what Yuval Noah Harari calls "dataism," which we encountered in chapter 3?

A UNIVERSAL MACHINE?

Life is a process, not a thing. In the words of Daniel Dennett, "It ain't the meat, it's the motion." Hofstadter attributes this remark to Dennett, saying "this was a somewhat subtle hat-tip to the title of a somewhat unsubtle, clearly erotic song written in 1951 by Lois Mann and Henry Glover, made famous many years later by singer Maria Muldaur."[13]

What is a process? It is simply change over time. A process is *discrete* if, at its essence, it is a sequence of separable steps. A process is not discrete if it is instead a fluid, continuous progression. For any physical process, whether it is discrete or continuous is a question of *modeling*. The same process can often be modeled as discrete or continuous.[14]

Consider water leaking into a boat. We can think of this as a fluid smoothly filling the boat, or as a torrent of individual molecules crossing a threshold from ocean to boat. But the essence of the concept of "water leaking into a boat" is better captured for most purposes by the continuous fluid model. In contrast, bailing a boat can be modeled as a discrete sequence of individual actions, filling the bailer and dumping it overboard, or as a continuous motion of an arm dipping the bailer and lifting and tilting it. But the essence of the concept of "bailing the boat" is better captured by the discrete model. The question, therefore, is whether a discrete or continuous model better represents the essence of the process.

Today's computing technology is all about processes that are better modeled as discrete. It is true that under the hood, digital electronics is just electrons sloshing around, no more a discrete process than water leaking into a boat. But the whole point of a computer, like a bailer for the boat, is to regulate the sloshing so that it is well modeled as being discrete. The physical processes that make up a computer are carefully designed to behave like sequences of separable steps, even if the underlying physics is continuous. Is biology similarly discrete? Some biological processes are well modeled as being discrete, but many are not. Are the ones that are not inessential?

Unfortunately, many thinkers today misunderstand a classic result, due to Alan Turing, and conclude that computers today are "universal machines," in principle. Specifically, they assume that a fast-enough computer with enough memory can do anything that any other machine can do. If this is true, then everything that any machine (including my brain) can do will eventually be doable by computers if we keep improving their speed and memory capacity. But Turing did not invent a universal machine. In fact, nobody has yet invented a universal machine.

A *Turing computation* is a discrete process that starts with well-defined digital inputs, carries out a discrete and finite sequence of steps governed by well-defined rules, and gives a final digital answer. Alan Turing's momentous contribution, published in 1936, was to describe a machine that can carry out any Turing computation, given enough time and memory. The set of all functions (mappings from inputs to outputs) that can be computed by a Turing computation are called the "effectively computable functions." This term implies that other functions are not effectively computable, but what it really means is that they are not computable by a Turing machine or any equivalent machine.

The machine Turing described, now called a "universal Turing machine" (see figure 8.1), has a digital input encoded on a tape, a program that is also digitally encoded on the tape, and a simple mechanism that reads and writes bits on the tape, moves the tape left and right, and switches between a finite number of states. The movements of the tape and the

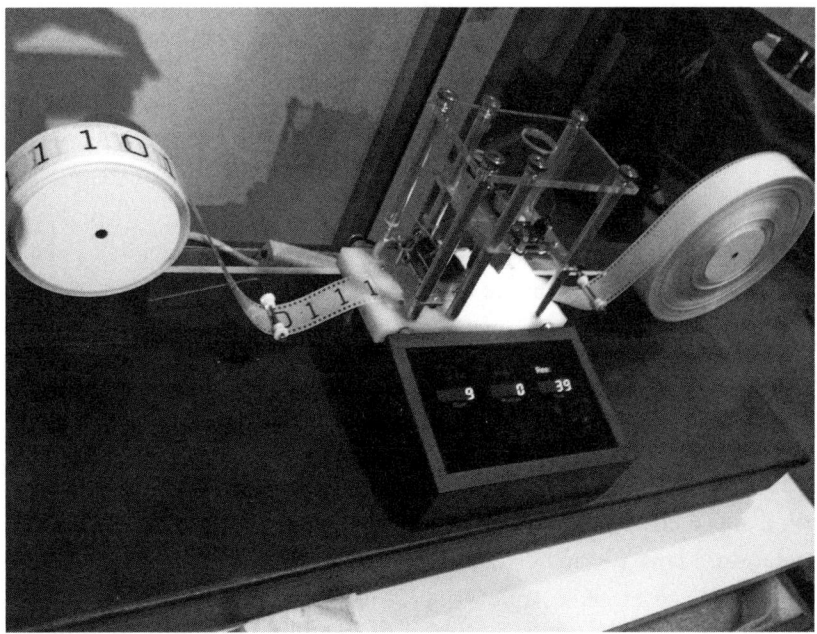

8.1 Machine designed by Mike Davey to resemble as closely as possible the hypo-thetical machine that Alan Turing described in his famous 1936 paper ("On Comput-able Numbers with an Application to the Entscheidungsproblem," *Proceedings of the London Mathematical Society* 42, pp. 230–265) as seen at the Go Ask Alice exhibit at the Harvard Collection of Historical Scientific Instruments. The builder's website is: http://aturingmachine.com. By GabrielF, CC BY-SA 3.0, via Wikimedia Commons.

reading and writing are governed by simple rules. One rule might be, for example, "if the current state of the machine is *A* and the value on the tape at the current read-write head position is 1, then move the tape one slot to the left and go to state *B*."

Although a universal Turing machine is not a universal machine, it is, nevertheless, an extraordinary machine. It can compute any effectively computable function, given enough tape and time, though on the face of it, that statement is tautological, since the "effectively computable functions" are defined to be those computable by a Turing machine. Not everything of interest is an effectively computable function, or even a function. First, a Turing computation has all its inputs available before it starts computing, so it cannot interact with the source of the inputs and affect the inputs in any way (see chapter 11 for a discussion of how inter-active processes are fundamentally richer than Turing machines). Second, the inputs must be digitally represented. Third, the state of the machine is digital and the mechanism has a finite number of possible states. Fourth, the operations of the machine follow a finite number of well-defined deterministic rules (it cannot, for example, flip a coin to decide which way to move the tape). Fifth, the output must be digital. Finally, the machine must terminate with an answer to have been deemed to have done anything at all. Each of these restrictions is rather severe, and we have little reason to believe that machines found in nature have any of these restrictions.

In principle, any Turing machine is realizable by a modern computer that has a sufficient amount of memory. This means that, given enough time and memory, any modern computer can perform any Turing computation. Many researchers have shown that universal Turing machines can be built on biological or chemical substrates,[15] but this does not imply that nature is limited to Turing computations.

Independently of Turing, also in 1936, Alonzo Church, an American mathematician, came up with a different model than the Turing machine that can also compute any effectively computable function. His machine was quite different from Turing's, and both are quite different from modern computers. It turns out that a wide variety of machine structures can all realize Turing computations, in principle. For computers, therefore, Hilary Putnam's multiple realizability (see chapter 7) is a reality. All these machines can perform exactly the same Turing computations, and the structure of the machine is incidental to what is happening.

What is now called the Church-Turing thesis states that every Turing computation, a step-by-step procedure (an algorithm) operating on digital information that terminates and gives an answer, can be computed by a Turing machine, and hence by any computer, given enough time and memory. Dataists frequently misrepresent the universal Turing machine, calling it a "universal machine" and stating that it can realize any other machine. Can it realize my dishwasher? My brain? They conveniently forget that a universal Turing machine can only perform Turing computations.[16]

Even computers can do things that a universal Turing machine cannot. Many applications, including Wikipedia and Google search, are designed to never terminate and are interactive (a property that I will return to in chapter 12). They are, by design, not Turing computations. They are therefore not realizable by the dataist's "universal machine."[17] Nevertheless, everything a modern computer does can be decomposed into a (possibly nonterminating) sequence of Turing computations. The biological world is probably not so limited.

BOREL'S AMAZING KNOW-IT-ALL NUMBER

Researchers have, over the years, come up with many elaborations of Turing computations, including ones that operate on nondigital data, can interact with their environment, or include nondeterministic operations. The term "hypercomputation" is used to refer to models of computation capable of computing functions that are not "effectively computable." Much of this work draws ire from faithful dataists, as this book is likely to do.

One such dataist is the mathematician and computer scientist Gregory Chaitin, who is well known for contributions to algorithmic information theory. Several hypercomputation concepts address computing with real numbers instead of digital data, something that seems to have prompted Chaitin to question the "reality" of real numbers.[18] He points out mathematical, philosophical, and computational difficulties with real numbers and concludes that these difficulties undermine the common assumption that real numbers underlie physical reality, concluding that physical reality is discrete, digital, and computational.

In a first example of the difficulties posed by real numbers, Chaitin cites the French mathematician Émile Borel, best known for his foundational work in measure theory and probability. Borel, in 1927, in Chaitin's words,

pointed out that if you really believe in the notion of a real number as an infinite sequence of digits 3.1415926..., then you could put all of human knowledge into a single real number.[19]

Chaitin calls this number "Borel's amazing know-it-all real number."

One way to construct Borel's number is to list, in some order, all the yes-no questions that have answers (Borel's questions were in French, and they could be ordered by length in characters and then alphabetically).[20] Then Borel's number can be represented in binary as $0.b_1b_2b_3 \cdots$, where b_i

is 0 if the answer to the i-th question is "no" and 1 if it is "yes." The resulting number is a real number between 0 and 1. Chaitin implies that Borel's number could not exist in any reasonable sense, presumably because it is not possible to "know it all."

There are some problems with this argument. First, what does it mean for a number to "exist"? The philosopher Immanuel Kant made the distinction between the world as it is, the thing-in-itself (*das Ding an sich* in German), and the phenomenal world, or the world as it appears to us. Let us assume that what Chaitin means by "exists" is that it is a thing-in-itself, in which case, whether the know-it-all number reveals itself to us, becoming a phenomenon, is irrelevant to its existence.

In fact, if Borel's number exists as a thing-in-itself, outside ourselves, then it *cannot* reveal itself to us. Shannon's channel capacity theorem ensures this. Borel's know-it-all number cannot be encoded with finite number of bits unless the list of all possible yes-no questions is finite, which it is not.[21] Unless we invent some noiseless way of observing a thing in itself, any observation will reveal only a finite number of bits. But this in no way undermines the existence of the number.

Returning to the original question, are real numbers real, I have to ask, are numbers real? There is real risk here of confusing the map with the territory. For numbers to be real, we have to assume a Platonic heaven where universal truths exist independent of humans. Numbers, be they whole numbers, rational numbers, or reals, would be premier citizens of such a heaven. Since that heaven's existence is independent of the existence of humans, then our knowledge of anything in it must be conveyed somehow to us through observation or through introspection. If it is conveyed to us through observation, then it will be subject to Shannon's channel capacity theorem, in which case we can only know about things that can be encoded in a finite number of bits. If it is conveyed to us through introspection, then its existence is a matter of faith, since its existence is independent of us, and there is no connection between that introspection and the thing-in-itself, by the definition of introspection. Either way, the difficulty is not with the number existing, but rather with our knowing the number.

A second example of the problems posed by real numbers is Richard's paradox, first stated by the French mathematician Jules Richard in a letter in 1905. Richard pointed out that all possible texts in French can be listed in some order in a manner similar to Borel's yes-no questions. A subset of these texts describes or names real numbers. But it is easy to describe a number that is not described on the list. Consider the phrase "the smallest number not describable in fewer than eleven words." These ten words seem to define a number that cannot be on the list of described numbers.[22] Every interpretation of a phrase in French or English as a number depends on the notion of semantics, the assignment of meaning to expressions. Semantics connects human cognition with the formal and countable world of symbols and sequences of symbols, but we have no evidence that the cognitive world is formal and countable.[23] Arguably, Richard's paradox demonstrates that written language *must* be ambiguous, perhaps because it bridges a countable world (phrases) with an uncountable one (meaning).

Let us assume that numbers are models (maps) reflective of some reality (territories). Under this assumption, Chaitin's question becomes one of whether real numbers are accurate models of some physical reality. We could ask the question of whether real numbers are *useful* models, but that question is trivial; we know they are. So let's focus on whether real numbers are *accurate* models of reality. Chaitin observes that some physicists argue that they are not:

The latest strong hints in the direction of discreteness come from quantum gravity…, in particular from the Bekenstein bound and the so-called "holographic principle." According to these ideas the amount of information in any physical system is bounded, i.e., is a finite number of 0/1 bits.[24]

As I have argued in chapter 8 of Lee, *Plato and the Nerd*, this "digital physics" hypothesis is not falsifiable, and therefore not scientific according to the philosophy of Karl Popper (more about this later). Digital physics can only be taken on faith.[25]

Chaitin also leverages biology to bolster his digital faith:

Other hints come from…molecular biology where DNA is the digital software for life…[26]

As pointed out by George Dyson in *Turing's Cathedral*,

the problem of self-reproduction is fundamentally a problem of communication, over a noisy channel, from one generation to the next.[27]

Since reproduction is a noisy channel, it can convey only information that can be encoded with a finite number of bits. DNA, therefore, might as well be encoded digitally. There would be no point in a richer encoding.

Chaitin's objection is fundamentally to the notion of a continuum, a mathematical concept that most certainly does lead to conceptual difficulties in the formal languages of logic and mathematics that humans have invented. But these formal languages live in a countable world, so it should not be surprising that they have difficulty comprehensively handling an uncountable world. Despite these difficulties, the cognitive notion of a continuum is not at all difficult to grasp. The difficulties arise only when trying to communicate, for example by naming or describing all the real numbers. But one can understand without communicating. In fact, conveying understanding is notoriously difficult. We call it "teaching."

Chaitin rests on the ancient Greeks when drawing his sweeping conclusion:

According to Pythagoras everything is number, and God is a mathematician. This point of view has worked pretty well throughout the development of modern science. However now a neo-Pythagorean doctrine is emerging, according to which everything is 0/1 bits, and the world is built entirely out of digital information. In other words, now everything is software, God is a computer programmer, not a mathematician, and the world is a giant information-processing system, a giant computer.[28]

This statement describes a faith, not a scientific principle.

TOO MUCH INFORMATION

If a DNA molecule fully defines a human, then we humans can be completely defined by a finite number of bits. Each rung of the DNA double-helix molecule (see figure 8.2) consists of a base pair, where each base is one of four types. Since there are a finite number of rungs and each rung is one of a finite number of types, the entire DNA molecule can be unambiguously identified by a finite number of bits.

In my own previous book, I questioned the completeness of DNA.

Only features that can be encoded with a finite number of bits can be passed from generation to generation, according to the channel capacity theorem. If the mind, or features of the mind such as knowledge, wisdom, and our sense of

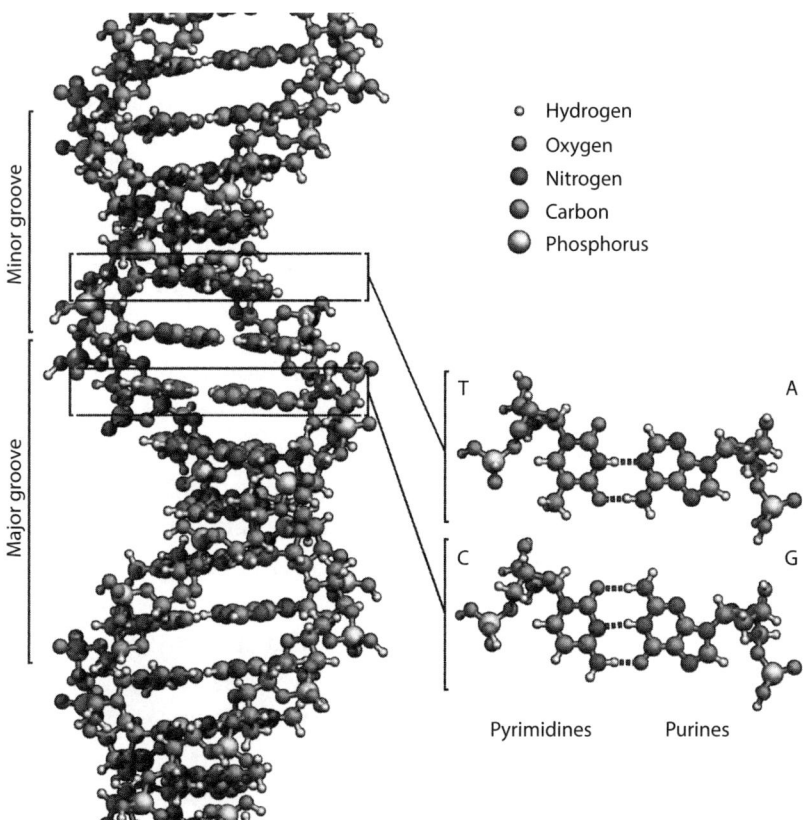

8.2 Structure of a segment of a DNA molecule, illustrated by Richard Wheeler (http://www.richardwheeler.net). By Zephyris, CC BY-SA 3.0, via Wikimedia Commons.

self, cannot be encoded with a finite number of bits, then these features cannot be inherited by our offspring. It certainly appears that DNA does not encode the mind because the mind of your offspring is not your own or even a combination of those of both biological parents.

If the mind requires mechanisms beyond digital for its operation and character, then the mind cannot be conveyed by any mechanism over a noisy channel. Your mind is entirely your own. Not only can it not be passed on to your offspring, it cannot be passed to anything. It will never reside in other hardware unless we invent a noiseless channel. Biological inheritance cannot provide a noiseless channel because if it did, there would be no mutation, there would be no evolution, there would be no humans, and we would have no minds at all. Genetic inheritance is, of necessity, digital, but minds are formed from more than genetics.[29]

Your DNA does not define *you*. Identical twins have (mostly) identical DNA but are different people. The dataist's dogma is that *you*, not just your DNA, can be fully defined by a finite number of bits. But in order to be able to distinguish identical twins as individuals, this is going to require more information than what is encoded in the DNA. How much more? Will any finite number of additional bits be sufficient?

Anything that can be fully defined by a finite number of bits can, in principle, be conveyed from one place to another at the speed of light, using a radio signal, assuming that some mechanism exists at the receiving end to reconstruct the physical thing from local materials using the bits as instructions. Amazingly, in principle, such a thing can be conveyed *perfectly*, without error, even though every radio channel suffers from random interference that engineers call "noise." In practice, there is always a chance of error, where some bit in a sequence of bits gets flipped in transmission. But the probability of such an error can be made as small as we like by just putting in more engineering effort. The theoretical limit is perfect communication, proved by Claude Shannon in 1948.

Under the assumption that all channels have noise, then any physical system that requires more than a finite number of bits to encode *cannot ever be copied*. The teleportation that Tegmark, Dennett, Parfit, and Hofstadter consider is such a copying. It can only possibly work if a cognitive human being can be encoded with a finite number of bits. A Turing machine can be encoded with a finite number of bits, so if a cognitive human being is a Turing machine, then teleportation is possible. But a cognitive human probably is not a Turing machine.

NOISELESS MEASUREMENTS

According to a highly influential principle articulated by the Austrian-British philosopher of science Karl Popper (1902–1994), the core of the scientific method is falsifiability. A theory or postulate is scientific only if it is falsifiable, according to Popper. To be falsifiable, at least the possibility of an empirical experiment that could disprove the theory must exist.

For example, the hypothesis that "all swans are white" is not *proven* by any number of observations of white swans. But observations of white swans do *support* the hypothesis; the hypothesis is falsifiable because an experiment *may* find a black swan. Hence, it is a scientific theory, albeit a false one.

Any empirical experiment must make measurements. The hypothesis that space and time are discrete is not falsifiable unless you assume that a perfect measurement apparatus, one whose measurements have exactly zero error, can be built. A measurement apparatus is a communication channel. It conveys information from the thing or process being measured to somewhere else, typically a computer. Every known measurement apparatus has noise, and therefore, every measurement reveals a finite number of bits of information. No such measurement, therefore, could falsify an assertion that a finite number of bits is required to represent the thing or process being measured. To falsify that assertion, the experiment would have to show that some aspect of the thing being measured requires an infinite number of bits to represent exactly, which means it would have to make a measurement the result of which requires an infinite number of bits.

According to Popper, for a postulate to be falsifiable, we only need for an experiment that falsifies it to be *possible*. We do not have to construct the apparatus for the experiment nor do we have to conduct the experiment. Therefore, digital physics could be saved, as a scientific theory, if a noiseless measurement apparatus is *possible*. Is it? I would conjecture that noiseless measurement is possible only if space and time are actually discrete. If I am right, then we have to assume that digital physics is true in order for digital physics to become a scientific theory. Otherwise, we can only take it on faith.[30]

IS TIME DISCRETE?

McCulloch and Pitts showed that neurons fire discretely, so it seems that the basic machinery of cognition has digital aspects. Moreover, the map of connections between neurons may be discrete, in the sense that whether one neuron is connected to another is just one bit of information, plus some bits to identify the neurons.

But what if it matters not only *whether* one neuron is connected to another, but also *where* the neurons are connected and *how well* they are connected? Recall the complicated brain structure revealed by Jeff

Lichtman discussed in chapter 3. At a minimum, the actual geometric layout of the connections will affect the *timing* of firings. This can change the order of events in the brain, which plausibly could lead to differing function. If this is the case, Lichtman's connectomics has an even more unachievable goal. A digital representation of the brain will become possible only if space itself is discrete. It will be possible only if the scientists who advocate digital physics happen to be right despite their flawed arguments. A digital representation of the current state of a working brain will become possible only if time is also itself discrete. That time is discrete is another questionable hypothesis of digital physics.

Although it remains controversial, many physicists do believe that time is discrete. Many of these dataist physicists cite Planck time, proposed by Max Planck, who won the 1918 Nobel Prize in physics for discovering that energy is quantized. The Planck time, which is approximately 5.4×10^{-44} seconds, is the time required for light to travel in a vacuum a distance of one Planck length.[31] The Planck length is approximately 1.6×10^{-35} meters. This length is twenty orders of magnitude smaller than the diameter of a proton, so these quantities are indeed truly tiny. Whether these minute quantities really represent a fundamental discretization of time and space remains unclear and probably depends on developing a working theory of quantum gravity, which remains elusive. As of this writing, there is no consensus among physicists about whether time or space are discrete. In the meantime, our best working models do not discretize time or space.

One of the strongest proponents of a dataist view of physics, John Archibald Wheeler, who coined the term "it from bit" to capture the digital and discrete nature of the physical world, wrote the following:

Time, among all concepts in the world of physics, puts up the greatest resistance to being dethroned from [the] ideal continuum to the world of the discrete, of information, of bits.[32]

There have been significant developments in physics since 1986, however, when he wrote this. More recently, the theoretical physicist Carlo Rovelli, who has made significant contributions to the field of quantum gravity, is more sure:

The "quantization" of time implies that almost all values of time *t do not exist.* If we could measure the duration of an interval with the most precise clock imaginable, we should find that the time measured takes only certain discrete, special values. It is not possible to think of duration as continuous. We must think of it as discontinuous: not as something that flows uniformly but as something that in a certain sense jumps, kangaroo-like, from one value to another.

In other words, a minimum interval of time exists. Below this, the notion of time does not exist—even in its most basic meaning.[33]

Note, however, that Rovelli is talking here about measurement, "if we could measure," and is concluding something about the thing in itself. By Shannon's channel capacity theorem, unless the measurement is noiseless, it will only reveal a finite number of bits of information, and consequently would inevitably make time *appear to be* quantized. But a measurement is not the thing in itself, and quantum theory, with its extensive use of nondeterminism, makes the prospect of noiseless measurement remote indeed. Rovelli says so himself:

The intrinsic quantum indeterminacy of things produces a blurring, ... which ensures—contrary to what classic physics seemed to indicate—that the unpredictability of the world is maintained even if it were possible to measure everything that is measurable.[34]

Nevertheless, Rovelli draws a sweeping conclusion, echoing that of Chaitin:

Continuity is only a mathematical technique for approximating very finely grained things. The world is subtly discrete, not continuous. The good Lord has not drawn the world with continuous lines: with a light hand, he has sketched it in dots, like the painter Georges Seurat.[35]

I believe Rovelli got this backward. The discrete, computational world approximates the continuous physical world.

In Rovelli's defense, he was the first proponent of what is called the relational interpretation of quantum mechanics, which states that the state of a quantum system is a relation between the observer and the system. That is, there is no notion of state that is independent of an observer. Under this interpretation, being able to observe only a finite number of bits means that there are only a finite number of bits. But this relational interpretation is not widely accepted even by proponents of digital physics.

Nevertheless, Rovelli makes an intriguing suggestion a few pages later:

The physical substratum that determines duration and physical intervals ... [is] a quantum entity that does not have determined values until it interacts with something else. When it does, the durations are granular and determinate only for that something with which it interacts; they remain indeterminate for the rest of the universe.[36]

Suppose that in physics *every* interaction is noisy, not just interactions that have the goal of measuring something. If this is the case, then every interaction can exchange only a finite number of bits of information, regardless of how much information exists in the essence. This would inevitably lead to the appearance of a quantized, digital world even if the world is not so. A natural question arises here. If some part of a physical system contains internally more information than can be represented with a finite number of bits, then that information cannot affect neighboring parts of the system because any interaction between parts must occur over a noisy channel. So, is it possible that the only relevant information is that which *can* be encoded with a finite number of bits? It seems that the answer has to be "yes" if the system is granular, a composition of discrete components. But if the system itself exists in a continuum, where one part smoothly blends into the neighboring parts, then the answer is not so clear. Interactions are no longer between discrete components over a noisy channel.

A more nuanced perspective on whether the world is discrete or continuous is given in *The Age of Intelligent Machines*, published in 1990 by the American futurist Ray Kurzweil. Kurzweil argued that whether a system is viewed as discrete or continuous depends on the level of abstraction at which one examines the system. Every discrete abstraction will have an underlying continuous mechanism, and below that continuous mechanism could lie a more fine-grained discrete mechanism. If Kurzweil is right, then not only is a discrete, computational model of the world incomplete, but so is a continuous one. Every model we construct of the physical world admits the possibility of an underlying, more fine-grained model that may have a different character.[37]

IMPERFECT COMMUNICATION

This book can be encoded with a finite number of bits. If you are reading a paper copy, if those even exist by the time you read this, then it is not the paper and ink that is encoded—it is the text and (digitized) images in the book. A paper book cannot be perfectly copied, but the text of the book can be.

Whether you, dear reader, are an artificial or a natural intelligence, the function of this book is to convey ideas from my cognitive self to yours. Do the *ideas* (versus the text) get copied perfectly? I doubt it. If I ask you questions about this book, the answers you give me will likely surprise me. If you are a natural intelligence, then I have no doubt that you understand my sentences, but that understanding is all yours, not mine. I'm pretty sure that these words do not and cannot actually convey the thoughts from my head fully, except to the extent that those thoughts take the form in my head of these words. I believe there is more to my thoughts than these words. And the frustration of writing is that even the most carefully crafted words will fail to adequately convey my thoughts.

Speech, as a sequence of words, is a digital encoding of thought. At any given time, in any given language, the vocabulary is finite. In written form, adding to the finite set of words a finite number of punctuation marks, everything said can be encoded digitally. Hence, according to Shannon, the words can, theoretically, be perfectly conveyed, even by a noisy channel. The *meaning*, of course, may change in transport, but the *message*, a text in a language, can arrive at the recipient unharmed.

A *spoken* sentence, compared to a written one, is a bit more complicated. As it happens, because of limitations in the human auditory system, the sound signal of a spoken sentence can, in fact, be perfectly reproduced by a finite sequence of bits. No human will be able to detect the difference between the original and the reproduction. But it takes many more bits to represent the sound signal than to represent the words. An audio book takes up much more space in your computer memory than a textual book. And indeed, an audio book conveys more information because the nuanced phrasing of the narrator doing the reading can convey meaning and interpretation that is not intrinsic in the words.

When we converse with someone over the telephone or over an audio-only Internet call, less information is conveyed than with a video call. A video call, in turn, conveys less information than an in-person encounter. Subtle changes in facial expression and body language affect the communication. A video signal can capture some of these effects, but not all of them. In particular, at least in today's technology, a video signal does not change with the movement of the viewer's head, for example. I cannot look around your back to see that you have crossed your fingers while lying to me through a video signal. Nevertheless, because of limitations in the human visual system, a finite number of bits is enough to essentially perfectly convey the image that a single, immobile human eye sees.

A video signal occupies even more space in a computer memory than an audio signal. The information carried is more, so we should expect to pay more to convey that information. When we combine text, audio, and video, we have gone beyond the core of language and are getting closer to real-world communication. But the nature of the communication has become quite a bit more complex. How much is a movie like the book it is based on? The ideas that are conveyed are now an amalgam of those in the minds of an author, a screenwriter, a director, an actor, a set designer, and so on. As I watch a movie, thoughts form in my head. Are those the thoughts of the author? Certainly not. My thoughts are all mine, pushed and nudged by the movie of course, but still all mine. Even a movie carries only a finite number of bits of information.

If my thoughts are not fundamentally digital, then not only do we lose information when conveying a message, we lose an *infinite amount* of information. We go from something that cannot be captured by any finite number of bits, a thought, to something that is perfectly representable with a finite number of bits, a text, an audio recording, or a video recording. No matter how high the resolution of your TV set and how expensive your audio system, the experience of watching the movie fundamentally cannot match the thoughts in the mind of the author. It cannot be done.

Accepting this, however, remains an act of faith. But there is considerable evidence for its validity. Think about the progression from reading text on a page, listening to a recording of the author reading it to you, watching a video of the author acting it out, watching a play where the author acts it out on stage, and finally, standing face-to-face with the author and having the author tell you, personally, what are her thoughts. Each step of this progression very likely makes your own thoughts better match those of the author. But you will never achieve a perfect match.

That your thoughts can never perfectly match those of the author stimulating those thoughts may seem frustrating, but perhaps it explains why we have a concept that we call "art." Art is fundamentally about better connecting human brains that can never be perfectly connected.

Consider reading a musical score, listening to a recording of the composer playing the piece, and being in the room with the composer playing the piece. Each step conveys more information and more deeply connects your brain with that of the composer. But the thoughts and feelings in your brain will not and cannot perfectly match those of the composer. Perhaps if thoughts were digital, then we would not have art.

The Lebanese-American poet, writer, artist, and philosopher Khalil Gibran wrote in 1923,

> Your children are not your children.
> They are the sons and daughters of Life's longing for itself.
> They come through you but not from you,
> And though they are with you yet they belong not to you.
> You may give them your love but not your thoughts,
> For they have their own thoughts.

Unless you have an identical twin or have been cloned, like Dolly the sheep (see figure 8.3), your children may be as physically alike to you as is possible, and yet they cannot house your thoughts. Even your twin or your clone cannot house your thoughts. How likely is it, really, that a hunk of silicon and metal can?

8.3 Taxidermied remains of Dolly, a clone created from a cell in the mammary gland of another sheep by Keith Campbell, Ian Wilmut, and colleagues at the Roslin Institute at the University of Edinburgh, Scotland, and the biotechnology company PPL Therapeutics. By Toni Barros, CC BY-SA 2.0, via Wikimedia Commons.

AH, TO BE DIGITAL!

As of this writing, any artificial intelligence in existence can be teleported, cloned, and backed up. By Shannon's channel capacity theorem, the same is true of our human selves only if we are digital or if noiseless measurements and noiseless communication are possible. I have argued that these are possible only if the physical world itself is digital, and that we can never know whether this is the case.

Just as we can never know whether the physical world is digital, we can never know whether teleportation actually works. If, as suggested by Tegmark, an apparatus were constructed that makes a copy of you, how can we be sure it is an *exact* copy? This would require making measurements that show that there are no differences. But unless these measurements are noiseless, they will reveal only a finite number of bits of information about both the original and the copy. Those bits may match, but in principle, there could be an infinite amount of additional information that the bits fail to capture. So the answer to the question, "Is life digital?," may be profoundly disappointing to many readers. The answer is, we don't know. We can't know. We can't ever know.

For machines, however, we know. Many of them are digital. It is possible that the most profound difference in this new life form on our planet, if they really are a life form, is that they are digital. This difference could account for some of the features that digital machines have that no biological living being on Earth has, and vice versa. In chapter 2, I pointed out that digital machines defined by software share bodies with one another. Your laptop computer, for example, is host to quite a few programs, and the hardware is the body for each. The Wikimedia Foundation servers host only a single program, Wikipedia, but servers maintained by Amazon Web Services, for example, can be hired out to form the "body" for any number of programs. Digital machines can even switch bodies while living. In cloud servers, tasks are often migrated to different servers to get better load balancing or to manage temperatures. These properties would be positively weird if they were found in biological life forms.

Digital technology is made differently from biological beings. Can it and will it nevertheless evolve human-like cognitive functions? This is the topic of the next chapter.

9

INTELLIGENCES

THE WRONG STUFF (AGAIN)

Even if you aren't convinced by the previous chapter, bear with me and assume for now that humans are neither digital nor algorithmic, whereas digital technologies are (mostly) both. Digital artifacts can be copied perfectly, teleported, and backed up. If they are living, then they can live forever. Our human selves have none of these properties.

We do, however, have a few properties that, at least so far, are not evident in any machine. We have a sense of self, a cognitive identity, and an ability to examine our own existence, for example by writing books like this one. We are intelligent (or at least some of us are), although frankly that statement is tautological because what we mean by "intelligent" is what human cognition does.

Although machines are pretty far from having a cognitive identity, they are not so far from exhibiting behaviors that we will be forced to classify as "intelligent." They can also exhibit charming forms of stupidity, demonstrated by the inane conversation between the Amazon Echo and Google Home in chapter 5. So perhaps being digital and algorithmic is not an impediment to achieving some of the cognitive properties of humans, both the good and the bad. If this is the case, then digital machines will have a huge advantage over us, in the long run, because of their ability to perfectly replicate and live forever. If this is the case, it is we humans who are made of the wrong stuff.

Perhaps biology will ultimately prove to be a mere stepping stone toward a more evolved digital life form. But there is another possibility. Being digital and computational is a *constraint*; it *limits* the possibilities. A technology that is not limited to being digital and computational can, fundamentally, do more than a technology that is so limited. Computers, after all, are ultimately built on an analog substrate. Transistors fundamentally operate in a fuzzy, continuous world. This analog world is obviously capable of realizing something digital and computational, since that's how computers are made, but it is also capable of other sorts of processes that are neither digital nor computational.[1]

Biology, like silicon, is capable of digital and algorithmic operations, as evidenced by DNA, the machinery of ribosomes, and the firing of neurons. But biology is not *limited* to those operations. It is capable of operations that no digital and algorithmic machine is capable of. If those operations are central to human cognitive functions, then digital and algorithmic machines will never replicate those functions.

But is replication of human functions really the goal? We are surrounded by machines that do things no human can do. Right now, as I write this, I am traveling at three hundred kilometers per hour on a bullet train from Beijing to Xi'an. This machine was not designed to replicate human functions. If it had been, then I would have climbed onto the back of a humanoid robot in Beijing, and the robot would run on two legs to Xi'an. Why do we insist on thinking that AIs will be cognitively anthropomorphic?

In the biological world, when two species share the same ecosystem, they often kill each other off or diverge into different niches. If they diverge into different niches, they are less likely to compete for resources. Kevin Kelly, whom we met in chapter 2, in his book, *The Inevitable*, argues that AIs will continue to exhibit distinctly nonhuman intelligences, as they do today.[2] Moreover, humanoid robots remain the stuff of science fiction and niche publicity stunts. When they get too much like humans, trying for example to emulate human facial expressions, the results can be creepy. The vast majority of successful robots do not physically resemble humans at all.

HUMANOID ROBOTS AND CREEPS

There is one good reason, at least, to fashion robots after humans, which is to make them more able to operate in a physical environment designed for humans. A nine-foot-tall robot, for example, will have difficulty getting through doorways, and a six-inch-tall one will have trouble climbing stairs. Stairs and doorways impose some constraints on scale, but they certainly do not require that robots have a face. Human morphology may be a legitimate *inspiration* for a robot design, but it's hard to see any reason for it to be a *requirement*, unless the real purpose of the robot is to be cute or to deceive us into believing that it is actually human. Robots on a factory floor do not sit in chairs, and self-driving cars are not realized by humanoid robots that wear seatbelts and press the accelerator pedal with their feet.

Machines that coexist with humans need to operate effectively in a world designed for humans. But that world will change with time. Back in the early 1980s, I worked at Bell Labs on the design of voiceband data modems (see figure 9.1). If you are old enough, you may remember these devices. They would connect your computer to the Internet through a telephone line. Today, your computer is constantly connected to the Internet, but back then, you would use your phone line to connect, and the phone line could not be used at the same time for voice calls. When you wanted to connect, your computer would tell the modem to dial a phone number, and the modem would produce a series of touch tones to dial the number. A modem at the other end would answer, and you could then hear an audible sequence of tones and squeals ending with a satisfying "shhhhhh-hhhh." You were then connected, and the modem would go silent.

The modem didn't really go silent. It just stopped playing its sounds over its speaker. If you were to pick up another extension on your phone line, you would hear the loud "shhhhhhhhhhh" continuing over the line. The job of the modem was to send a sequence of bits over a channel that was designed to carry human voices. Claude Shannon, whom we encountered in chapter 8, showed that the most efficient sound that accomplishes this sounds like white noise, which sounds like "shhhhhhhhhhh." This is a sound that contains all the frequencies that the phone line can carry in equal amounts.

9.1 A vintage voiceband data modem, the first to be realized by software, designed by the author and colleagues at Bell Labs in the early 1980s.

A more humanoid design of a voiceband data modem would be one that would speak over the phone line the sequence of bits that were to be transmitted. If you had such a modem, when you were to pick up another extension of the line while it was operating, you would hear "one, zero, zero, one, one, one, zero..." After all, the telephone channel was designed to carry human voices, so why not have the modem emulate a human voice?

But this would be a rather silly design. The modem that I designed sent 2,400 bits per second using a sound that sounded like "shhhhhhhh." To match 2,400 bits per second, the humanoid speaking modem would have to speak so quickly that no human would be able to understand anyway, so there would be little point in doing this.

Your computer today is probably capable of connecting to the Internet at more than a billion bits per second. The network is no longer one designed to carry human voices. On the contrary, today, when you make a voice call, your voice is converted to bits and sent over a network that is designed now to carry bits, not voices. The human-centric environment evolved to better accommodate the machines. I suspect we will see much more of such evolution in the future. Highways will change to better serve self-driving cars and cities will change to replace retail with deliveries and parking with pickup and drop-off zones.

The voiceband data modem was designed to operate in a human environment, but not to interact directly with humans. The most dramatic breakthroughs in AI in recent years have been in classifying images, understanding spoken natural language, and synthesizing natural language. These capabilities make computers more humanoid; they are designed not only to operate in a human environment, but also to interact with humans. A humanoid robot that needs to climb stairs need not adopt human morphology to do so efficiently, but if it is to be a reassuring companion for a human, perhaps it will need at least some human-like features, like Asimo, the robot we encountered in chapter 7 (see figure 7.2).

Robots that come too close to human morphology, however, are positively creepy. Sophia, a robot created in 2015 by the Hong Kong–based company Hanson Robotics, has a face modeled after actress Audrey Hepburn with eyes that blink, a mouth that smiles, and eyebrows that move, all on a head that moves and tilts in an attempt to emulate human gestures while conversing with a human (see figure 9.2). She can display some fifty facial expressions. In October 2017, Sophia was granted citizenship of Saudi Arabia, occupying what had been until that time a uniquely human niche. Watching her in action, however, is unnerving. Her gestures and facial expressions do resemble those of a human, but a very quirky human.[3] No doubt the technology for emulating human facial expressions can be improved, but do we want this?

TONE-DEAF AIS

In May 2018 Google demonstrated an AI called Duplex that took the idea of a personal digital assistant to a new level. You can ask this AI to make phone calls for you, for example to make an appointment for a haircut.

9.2 Sophia, a robot designed by Hanson Robotics. By ITU Pictures, CC BY 2.0, via Wikimedia Commons.

It makes the call, and, with a natural-sounding synthesized voice, negotiates with whoever answers the phone without revealing that it is a robot. Google put a great deal of effort into making the voice pass for human, using quirks of natural language like "um," responding with "mh-hmm" instead of "yes," and using a colloquial intonation called "upspeak," where the pitch rises in a statement so it sounds like a question. In a demonstration at their annual developer conference, Google demonstrated Duplex interacting with a confused restaurant worker who wasn't really listening. Duplex kept up in an astonishingly human-like way, concluding with "I gotcha" when it became clear that a reservation wasn't really needed for four people at this particularly restaurant on a Wednesday night. The AI appeared to be able to genuinely pass the Turing test, at least within the narrow domain of making reservations and appointments.[4]

The Google researchers were caught by surprise by the vehement public outcry denouncing the work. In a report on May 24, 2018, CNET said that the core feature of Duplex is "a deception" and asked, "why are we making technology to trick humans?"[5] The project has been called "tone deaf," and there have been calls for regulation to require such AIs to reveal that they are not human. Perhaps they could introduce themselves as follows:

Hello, I am Duplex. You can call me Dupe for short. I am Joe Schmoe's digital assistant, and I am calling to, um, make an appointment for Joe to have a massage.

Given that regulations have been completely ineffective at stemming sleazy robocall operations, there is no reason to believe that regulations requiring these machines to identify themselves as robots will be effective either. Imagine putting this in the hands of Russian political saboteurs and other political activists.

Hello, I'm Dupe. I, um, think we've met at church a couple times. Maybe you 'member me? You do? Great! Well, you may have heard that the feds are, um, planning a big military exercise up the road a piece at Camp Swift. Ya know the place, right? Well, I've heard that they are, um, secretly planning to round up honest folk who don't agree with this government. This is kinda scary, yeah, you bet....Anyway, we're tryin' to get folks who don't want martial law to call the governor's office...

I didn't (couldn't) make this up, or at least not completely. In May 2018, the former CIA director, Michael Hayden, revealed that Russians had used social media to spread conspiracy theories about a routine military exercise called Jade Helm that the Obama administration ordered in 2015.[6] In reaction to the ensuing hysteria, Greg Abbott, the Republican governor of Texas at the time, ordered the Texas National Guard to monitor the actions of the US military, an unprecedented move. The Russians managed to fuel the hysteria with technology much more primitive than Google's Duplex. Imagine what they can do now.

I'm afraid that the genie is out of the bottle; there is no rolling back. You cannot trust a voice on the phone unless you recognize the voice. But Google will soon be able to fake the sound of your voice too. Perhaps at next year's developers conference they will introduce "MyDupe, the digital assistant with your own voice!" A Canadian startup called Lyrebird (I'm not making that up either) already claims that they can replicate your voice given only a one-minute sample.[7]

The next obvious step is to have MyDupe call your friend's friend that you met last night to ask for a date. Using the robot instead of calling yourself will spare you the personal embarrassment of rejection. For an extra fee, you could purchase the charm module to improve your chances. Of course, once this has been accomplished, then the next step is to have MyDupe *answer* the phone too. Pretty soon, all voice communication will be robots talking to robots, like the Echo and the Home's endless babble in chapter 5.

Facebook, among others, has experimented with AIs talking to AIs. In one experiment that received quite a lot of media coverage in 2017, Facebook tested chatbots that negotiated with each other over the ownership of virtual objects.[8] The bots were programmed to experiment with language to try to dominate the negotiation. After a few days of learning from each other, the bots seem to have developed a language of their own:

BOB: I can can I I everything else

ALICE: Balls have zero to me to me to me to me to me to me to me to me to"

Oddly, they were able to conclude negotiations over ownership using this mysterious morphing of the English language. Various media sites raised the alarm, suggesting that the bots were developing their own language to elude their human masters. More likely, this tells us something about how language can evolve in an isolated setting, where entities communicate between themselves and not with outsiders. Facebook killed the project, not because it was embarrassing, but because they were interested in bots that could communicate with humans.

To reliably pass the Turing test with humans in more varied contexts will require endowing AIs with human traits that are not uniformly appealing. They will need to be unpredictable, capricious, petty, greedy, and devious. Without these traits, it will be easy to tell that they are not human. Once they achieve this level, however, they will be competing with humans in the niches of the same cognitive ecosystem. Perhaps your MyDupe will actually develop a romantic relationship with your friend's friend's MyDupe while you stay home and binge watch NCIS alone.

In a biological ecosystem, when two species compete for the same niche, they rarely successfully coexist. No other hominid species has survived in competition with *Homo sapiens*, to name a rather prominent example. Are the machines evolving into the same niche as us in our ecosystem? Martin Ford, whom we encountered in chapter 3, believes so. His book, *Rise of the Robots*, pits the machines directly against the humans in the same economic niches.[9] His thesis is that the machines will systematically replace humans in many working roles, sidelining us and leaving us idle.

We *Homo sapiens*, however, did not come to completely dominate the planet by quietly succumbing to competition from other creatures. Instead, we became who we are through a complex process of speciation, where populations diverge and become distinct species; competition, where one species displaces another, sometimes brutally; and hybridization, where distinct populations with divergent genetic features interbreed (see figure 9.3). If the machines were to try to obliterate us like we did to the woolly mammoth, we will almost certainly fight back. And humans have an impressive track record of fighting and killing the competition. The machines will likely get their comeuppance, but I doubt we will simply obliterate them. They are too valuable to us. But any speciation within the machines that leads to hostilities will likely be met

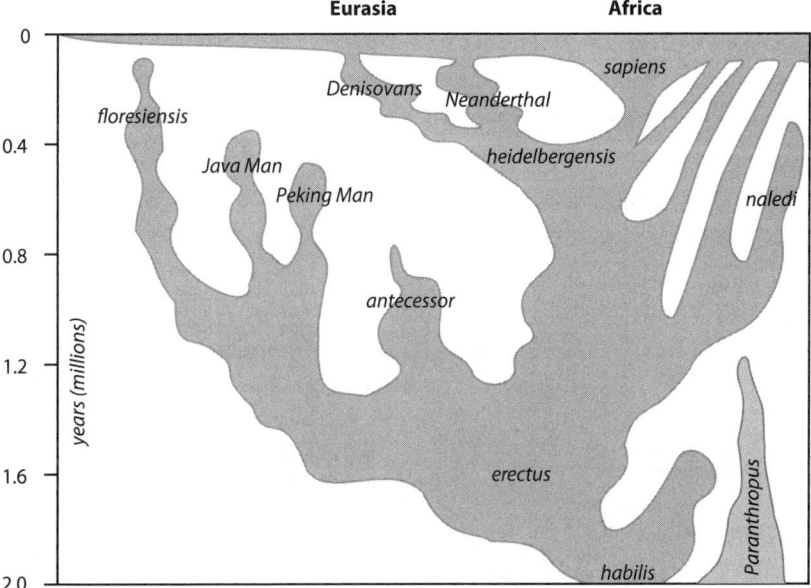

9.3 Sketch of speciation and hybridization of the genus *Homo* over the last two million years. From original by User:Conquistador updated by User:Dbachmann, CC BY-SA 4.0, via Wikimedia Commons.

with vigorous resistance or possibly a morphing of the ecosystem. Dupe will become ineffective either because voice communication will become obsolete or because we will find some way to effectively apply a death sentence to any duplicitous machine, treating it as a criminal.

TRANSHUMANISM AND THE SINGULARITY

If we assume the embodied cognition perspective described in chapter 8, then machines and humans are already undergoing a form of hybridization, where our cognitive selves, though not yet our genetic biological selves, already include some machine heritage. We have arguably already made them part of us. So Musk's "existential threat to humanity" (see chapter 3) may be real, but not in the sense that we will cease to exist. We just won't ever be the same again.

An occasionally odd and cultish movement called "transhumanism" vigorously embraces this transformation. Nick Bostrom writes about the historical roots of transhumanism, saying that humans have long had a "quest to transcend our natural confines."[10] He gives a tour of historical forms of this quest, such as the search for immortality, for example, the Fountain of Youth, as well as attempts to engineer our genetics through eugenics and the Nazi movement. Bostrom argues that AI, the singularity, nanotechnology, and uploading (the transfer of a human mind to a computer) are simply more modern versions of this quest.

Bostrom credits the statistician Irving John Good with the first clear articulation of a cognitive form of transhumanism:[11]

Let an ultraintelligent machine be defined as a machine that can far surpass all the intellectual activities of any man however clever. Since the design of machines is one of these intellectual activities, an ultraintelligent machine could design even better machines; there would then unquestionably be an "intelligence explosion," and the intelligence of man would be left far behind. ... Thus the first ultra-intelligent machine is the *last* invention that man need ever make, provided that the machine is docile enough to tell us how to keep it under control.[12]

But the design of machines may not be purely an "intellectual activity," but may actually be more the result of a Darwinian evolution, or more precisely, a coevolution with the human culture with which the machines are so intertwined (see chapter 14). If this is the case, it is not so clear that such an intelligence explosion is inevitable, nor is it clear that it will leave humans behind, since we too will evolve. Nevertheless, the idea has caught on, gaining an almost unquestioned currency among some people. The inflection point at which this intelligence explosion is to take off was called the "singularity" by Vernor Vinge, whose influential 1993 paper, "Technological Singularity," predicted that within thirty years, we will have entered a "posthuman era" that could even lead to the physical extinction of the human race.[13]

Emphasizing the link between humans and machines, my colleague at Berkeley, Ken Goldberg, has suggested replacing the term "singularity" with "multiplicity," representing "diverse groups of people and machines working together to solve problems." He points out that we already have this multiplicity:

> Multiplicity is not science fiction. A combination of machine learning, the wisdom of crowds, and cloud computing already underlies tasks Americans perform every day: searching for documents, filtering spam emails, translating between languages, finding news and movies, navigating maps, and organizing photos and videos.[14]

All of these technologies have a heavy human hand not only in their creation and development, but even in their very functioning. Google Search, for example, learns from each human-driven search to improve future search results for other humans.

Whether it takes the form of a singularity or a multiplicity, transhumanism, using technology to transcend our limitations, has been a reality for some time. Consider shoes and eyeglasses, for example. Cognitive transhumanism is also already a reality. My smartphone remembers many things for me and the spreadsheet in my computer does calculations far more reliably than I can. Do these technologies forebode extinction? This is quite a stretch, in my opinion.

The prospect of AIs completely sidelining humans, making us irrelevant, really doesn't make sense from an evolutionary perspective. Any significant threat to human well-being will be met with counterattacks, but more importantly, with adaptation. We are already adapting. Consider for example Ford's prediction that robots spell the end of work.[15] Far from the economically disastrous threat that Ford envisions, today's rapid technological development appears to be, as of this writing, coinciding with significant economic growth in much of the world. Ford's prediction that "this time is different," that automation will kill off many prospects for human employment, has shown some signs of materializing, but the story is complex, and the data is subject to conflicting interpretations. Despite the astonishing acceleration in automation, employment of humans is at record levels, with many industries suffering from an insufficient labor force, at least in more developed countries. Of course, this situation may have changed dramatically by the time you read this book. I am *sure* the situation will have changed, but I'm not so sure that it will be for the worse.

Many of the doomsayers seem to model our ecosystem in a rather simple way, where there is only one niche for intelligence. In such an ecosystem, it indeed seems likely that the smartest species will win. But what if the ecosystem is more complex and intelligence cannot be measured on a linear scale? IQ tests, as discussed in chapter 3, attempt to put intelligence on a linear scale. Given two agents A and B, there are only three possibilities: A is smarter than B, B is smarter than A, or they are equally intelligent. This is not really true even within the single species of humans, so it's hard to see why it should be true across species. A more common scenario is that A is better than B at cognitive function X and B is better than A at cognitive function Y.

The same will be true even within the world of machines. We will see not a monolithic superintelligence that subsumes all others, but a diversity of intelligences, many of them new and unfamiliar, each occupying a different niche in a complex ecosystem. Computers, even if they are made of the wrong stuff, are already better than any human at certain cognitive functions, such as finding relationships in very large bodies of unorganized information. Arguably, they have already developed some forms of intelligence that are distinctly not human. Much of the debate about whether these behaviors are "intelligent" stems from a history where the only intelligent agents in existence were human. The intelligences that are most likely to survive are the ones that complement rather than duplicate the capabilities of humans and of each other.

GOALS, ADAPTABILITY, AND A MISWIRED THERMOSTAT

IQ tests give us an imperfect way to measure intelligence in humans. I suspect that a machine trained for the purpose could possibly do well on such tests, but most AIs would fail abysmally. Does this mean that those machines possess no meaningful intelligence? Or does it just mean that IQ tests can only measure the intelligence of humans?

Shane Legg, a co-founder of Deep Mind, which was acquired by Google in 2014, and Marcus Hutter, a professor of computer science at the Australian National University, say that "intelligence measures an agent's ability to achieve goals in a wide range of environments."[16] Based on earlier work by Hutter in 2004, they give a model that, in effect, measures this form of intelligence. No IQ test needed. Their key idea is that merely achieving goals is not sufficient to be considered intelligent. If it were, then a thermostat would be intelligent. One must be able to achieve goals over a wide range of environments. This measure of intelligence is more flexible than IQ tests because it leaves more open as to what the goals might be.

However, it is tricky to define such a measure of intelligence because, in principle, the range of possible environments is vast indeed, and even humans would not do well in most possible environments. Neither a thermostat nor a human will be able to achieve any goals at all on the surface of the sun, for example.

Legg and Hutter assume that environments occur at random according to some probability, and they define intelligence as the average degree to which goals are met in a given environment weighted by the probability of that environment occurring. In other words, an intelligent being need not do as well in an unlikely environment as in a likely environment. Since thermostats are not likely to find themselves on the surface of the sun, perhaps they are intelligent after all.

A key challenge is to find some way to assign probabilities to possible environments. Legg and Hutter assume that each environment can be modeled as a computable function, and they assign a probability proportional to the complexity involved in computing that function. This means that an agent that fails to achieve its goals in a simple environment (one that is easy to compute) is penalized more than an agent that fails to achieve its goals in a complex environment (one that is difficult to compute). How difficult is the surface of the sun to compute compared to your living room?

For a humble thermostat, its environment is the room. Unfortunately, the thermodynamics of a room, based on widely accepted models in physics, is not actually computable, rendering Legg and Hutter's formalism technically inapplicable. The reaction of the room to turning on the heater can be modeled by partial differential equations that will exhibit chaotic behavior due to turbulent airflow. These differential equations operate in a time and space continuum, which is beyond the skill set of any Turing-Church computation, and they exhibit chaos, meaning that any computable approximation will fail to predict the behavior except over a short time horizon.

A similar form of chaos is exploited by San Francisco's Cloudfare, a tech company responsible for the security of a large number of high-profile websites. Cloudfare aims a camera at a wall of lava lamps, those 1960s-era psychedelic blob tanks, to generate encryption keys that are less predictable than those generated by any computation.[17] Like the room full of air affected by the thermostat, a lava lamp is a chaotic convection-driven bubble of fluids. As a consequence, our humble thermostat achieves its goals (keeping temperatures close to its set point) in an environment that is astonishingly complex, having infinite complexity according to Legg and Hutter's model.

Nevertheless, we can leverage the intuition behind Legg and Hutter's model to conclude that the thermostat is not as intelligent as a human being. Suppose that we modify the environment so that whenever the thermostat turns on the heater, the miswired room turns on an air conditioner instead. When the thermostat acts to turn on the heater, the temperature in the room will drop rather than rise. The thermostat would quickly saturate, running the air conditioner all the time, the room would get cold, and the electric bill would skyrocket. (I actually stayed in a hotel room once that seemed to work this way.)

The miswired environment is arguably no more complex than a correctly wired environment. Although both have infinite complexity by Legg and Hutter's measure, the equations describing the thermodynamics are nearly the same. If we consider these two environments to have equal complexity, then they would be equally probable by Legg and Hutter's measure, and the humble thermostat would do well in one environment, but fail abysmally in an equally probable environment. The humble thermostat can only deal with a narrow range of environments, those with a properly working heater in the thermodynamic environment of a room.

How would a human react in this environment? A human would quickly detect that the heating system is not working properly and would very likely shut it off. This would come closer to the goals than the thermostat would because it would prevent the room from continuing to get colder. Hence, if we were to extend Legg and Hutter's measure to allow noncomputable environments, we would have to conclude that the human is more intelligent than the thermostat. The thermostat is dumb because it continues to do what it does despite evidence that what it is doing is not working. The environment is not as expected, and the thermostat has failed to exhibit adequate adaptability. Moreover, the amount of intelligence exhibited by humans in this case is substantial, exceeding what any other animal would accomplish. Turning off the system is a uniquely human reaction that even the most sophisticated "smart" thermostat today would probably not do, yet almost any human would.

What the thermostat lacks is secondary, tertiary, and higher levels of feedback, as discussed in chapter 5. It is these higher levels of feedback that would endow the system with a form of "self-awareness," a recognition not only that it should act in a certain way if the room is colder than it should be, but also that it is its own actions that are expected to correct the situation. When these actions fail to correct the situation, the error signal in the higher-level feedback becomes large, demanding corrective action.

What Legg and Hutter have accomplished with their model is to give a measure of intelligence that does not depend on the metaphysical status of the agent but rather depends only on the effectiveness of its behavior. In the words of Stuart Armstrong, a research fellow at the Future of Humanity Institute at Oxford University, "a being is intelligent if it acts in a certain way."[18] Armstrong puts aside the philosophical problems with intelligence, saying,

For some, it can be fascinating to debate whether AIs would ever be truly conscious, whether they could be self-aware, and what rights we should or shouldn't grant them. But when considering AIs as a risk to humanity, we need to worry not about what they would be, but instead about what they could do.[19]

A key feature of intelligence, therefore, is how it actively adapts to a variety of environments. Humans react well to a more diverse suite of environmental conditions than any other being on Earth today, including all existing AIs. Is this situation temporary? It is certainly possible to design a smarter thermostat that will stop wasting energy if its actions are counterproductive. The real breakthrough, however, will be to design a thermostat that will figure out on its own to do this, without having to be explicitly programmed. This is very likely achievable with more layers of feedback loops.

WHAT DO YOU KNOW?

Most people know that one key feature of intelligence is knowing things. But what does it mean to know something? Does Wikipedia "know" anything? Everything? The question of what it means to "know" is, of course, the ancient philosophical problem of epistemology, or to use a friendlier term, theory of knowledge. The ancients, however, had not encountered machines like Wikipedia that seem to "know" a great deal. How does the existence of these machines change this ancient problem?

The English philosopher John Randolph Lucas, in an influential paper entitled, "Minds, Machines, and Gödel," offered some insights about what might be different between "knowing" in a human sense and "knowing" in a computer:

> In saying that a conscious being knows something we are saying not only that he knows it, but that he knows that he knows it, and that he knows that he knows that he knows it, and so on, as long as we care to pose the question: there is, we recognize, an infinity here, but it is not an infinite regress in the bad sense, for it is the questions that peter out, as being pointless, rather than the answers. The questions are felt to be pointless because the concept contains within itself the idea of being able to go on answering such questions indefinitely. Although conscious beings have the power of going on, we do not wish to exhibit this simply as a succession of tasks they are able to perform, nor do we see the mind as an infinite sequence of selves and super-selves and super-super-selves. Rather, we insist that a conscious being is a unity....[20]

Lucas implies that were we machines, we would get stuck in an infinite loop with each "knowing," performing an infinite "succession of tasks" or construing the mind as an "infinite *sequence* of selves." In this paper, Lucas goes much further and argues that "minds cannot be explained as machines" and that this fact is proved by Gödel's incompleteness theorems. But I believe that Lucas has defined "machines" too narrowly here. The key is his focus on *sequences* and *successions*.

In chapter 5, we saw feedback loops that were not made of sequences nor successions of discrete, separable steps. The feedback loops of Harold Black at Bell Labs and Norbert Wiener at MIT had no such steps. They were not algorithmic, and yet they were most certainly machines. Feedback loops in the setting of a continuum do not get stuck in an infinite regress despite being infinitely self-referential (uncountably infinitely, even). Lucas's argument may not apply to machines in general, but it definitely applies to computers, as we know them today, which are definitively algorithmic. Everything a computer does, it does as a sequence of discrete, separable steps.

To fully understand Lucas's argument requires understanding Gödel's incompleteness theorems, which Kurt Gödel published in 1931 when he was only twenty-five years old. His theorems put an end to a decades-long effort known as Hilbert's Program. The eminent German mathematician David Hilbert, around the turn of the twentieth century, tried and failed to put mathematics on a sound foundation as a formal language.

Despite their enormous importance, I will not rehash Gödel's theorems here,[21] but I will simply point out that the bedrock of a formal language is the notion of proof, which is a sequence of separable, discrete steps that demonstrate the truth of some assertion. The incompleteness theorems apply only to systems of reasoning that involve only discrete, separable steps. They do not apply to the forms of feedback investigated by Black and Wiener. My conjecture is that the human form of knowledge more closely resembles those latter feedback loops than the self-referential statements of Gödel.

Lucas actually comes strikingly close to the same conclusion:

The paradoxes of consciousness arise because a conscious being can be aware of itself, as well as of other things, and yet cannot really be construed as being divisible into parts.[22]

Douglas Hofstadter, in a 1982 article in *Scientific American*, finds flaws with Lucas's reasoning, but nevertheless appreciates his central point:

[Lucas] correctly observes that the degree of nonmechanicalness one perceives in a conscious being is directly related to its ability to self-watch in ever more exquisite ways.[23]

The feedback loops of Black and Wiener self-watch in ways that no computer can, despite being machines.

Hofstadter sees consciousness as a tangle of two goals, "flexible perception" and "self-watching," and observes that

There is no chronological priority here, because the two goals are too intertwined for the one to precede the other. It is a tricky fold back....

It is the trickiness of this "fold back," I believe, that inspired the title of his 2007 book, *I Am a Strange Loop*. Without chronological priority, this mechanism cannot possibly be algorithmic. But I see no reason that it cannot be mechanical (or electrical, hydraulic, chemical, or biological). It is out of reach by computers but not by machines. I believe that my brain is such a machine.

THE HARD PROBLEM

There are machines that do things that no computer can do, so even if computers cannot achieve consciousness, we cannot conclude that *machines* cannot achieve consciousness. Perhaps there are hints of how this might work in ever more exquisite feedback loops, but these are just hints. We are still fairly clueless. Hofstadter's former PhD student, David Chalmers, who appeared in chapter 7 as co-author with Andy Clark of the term "cognitive extension," calls consciousness the "hard problem." He argues for considering consciousness to be an "an irreducible entity (similar to such physical properties as time, mass, and space) that exists at a fundamental level and cannot be understood as the sum of its parts."[24] He is saying that no amount of advance in biology and neuroscience will explain consciousness.

Chalmers is in good company with his skepticism about how much current science can help. The eminent English physicist, mathematician, and philosopher Sir Roger Penrose, in his controversial book, *The Emperor's New Mind*, argues that while consciousness is a naturally occurring process in the physical world, it is not explainable using the known laws of physics, and hence certainly not explainable as a computation. The title of his book takes direct aim at the AI agenda of the time, to replicate human intelligence in computers. He says,

The belief seems to be widespread that, indeed, "everything is a digital computer." It is my intention, in this book, to try to show why, and perhaps how, this need not be the case.[25]

He goes much further to say that not only is consciousness not a computation, but that we are going to need a new understanding of physics, perhaps related to quantum gravity, before we will understand how the physical world can give rise to consciousness.[26] If Lucas, Chalmers, and Penrose are right, then no machine whose brain is a digital computer will ever achieve consciousness, no matter how far the technology advances.

CAN YOU LEARN IF YOU CAN'T KNOW?

The most dramatic successes in AI recently have been in the field of machine learning. Once something is learned, doesn't it become known? If you can't know, can you learn?

Machine learning today is more like acquiring the ability to perceive than like acquisition of knowledge. Consider the ability that computers have today to convert speech into text. The machine "knows" some words and learns to convert sound into those words, but even after learning, it does not, in the usual sense of "knowing," know how to convert sounds into words. Our brains do the same thing: our ears sense sound and the brain converts it to words. But we do not really "know" how to convert sound into words. Our brains just do it. We cannot teach it to another person and we cannot explain how it works. So, arguably, this isn't really "knowledge."

Before the recent successes of deep learning, a small community of experts in speech recognition believed that they possessed this knowledge, or could possess it, given enough research funding. Speech recognition was accomplished in computers by careful design of elaborate signal processing algorithms. Engineers would form a hypothesis about how sound could be converted into words, developing fancy techniques such as linear-predictive coding, hidden Markov models, and dynamic time warping. They would then "teach" the computers these algorithms by writing programs. But these speech recognizers never got very good. At best, they could reliably distinguish words in a small alphabet, such as "yes" and "no," and the ten digits.

Deep learning changed all that. We now teach machines how to recognize speech by feeding them examples. Given a sound, the machine converts it to text, doing very badly at first. The machine compares its guess against the correct text and then applies the backpropagation algorithm to fine tune its parameters so that on the next sound it does ever-so-slightly better (see chapter 4). This is a form of learning by doing, and it relies heavily on feedback.

The "doing" here, however, is of a rather primitive form. The only thing these machine learning algorithms are "doing" is classifying sounds. What if they could also make the sounds? Would this improve the way they learn? Perhaps they would need far fewer examples. Recall from chapter 5 the role of efference copies in speech production and understanding in the brain. Perhaps with such more elaborate feedback loops, digital machines will get better at learning.

In children, learning to recognize written characters is accelerated by drawing characters on paper. Older learners, those who already have considerable experience manipulating the physical world, may be able to *imagine* drawing the characters on paper instead of actually drawing them on paper. But even in older learners, actually drawing the characters on paper seems to help. I have personal experience with this because, although I have never studied Chinese, I have learned to recognize a small set of Chinese characters. To learn them, however, I found that I had to reproduce them on paper, even though I am an older learner. For me, imagination was not enough. When my kids studied Chinese in school, their instructor insisted on tedious exercises of filling pages with repeated drawings of a character. Given how tedious this is, I have to assume that if it were not effective, it would not have survived as a teaching method.

The cognitive scientist Josh Tenenbaum, working at MIT, believes that AIs today are not adequately exploiting this principle. He points to the fact that deep learning systems, like those considered in chapter 4, require many training examples of images of a hand-written character before they can reliably recognize it, whereas humans can often see just one example in an unfamiliar alphabet and be able to pick out that character, even with significant variations in its rendering. It may well be that humans encode the *concept* of the character more by the motor efferences that would result in the character being put to paper than by some representation of the visual pattern. The concept of the character, under this hypothesis, is tied tightly to the feedback loop of using our muscles to produce a pattern on paper and seeing that pattern emerge in our visual field. Could learning to read be so tightly bound to learning to write?

Working with Brenden Lake (then a graduate student and now on the faculty of psychology and data science at New York University) and postdoc Russ Salakhutdinov (now on the faculty at Carnegie Mellon University), Tenenbaum demonstrated a software system that represents characters in an unfamiliar alphabet as programs that can draw the character.[27] Given a single hand-written instance of a new character in a new alphabet, their program was able to outperform deep neural nets at classifying the character in another handwritten text.

Even more interesting, the program created by Lake, Salakhutdinov, and Tenenbaum was able to produce convincingly handwritten versions of a new character. In their 2015 paper in *Science*, they report showing their program a single instance of an unfamiliar character and asking it to produce novel instances that are similar. They presented humans, recruited through Amazon's Mechanical Turk service, with the same task, and then asked judges to determine which characters were created by machines and which by people. The characters produced by their program were convincingly handwritten, consistently fooling the judges. Their program, therefore, within this narrow domain of handwriting, passed the Turing test. Compare this to the bizarre images synthesized by DeepDream, discussed in chapter 4, which no human would confuse for the real thing. Recall from chapter 4 that generative adversarial networks were able to synthesize better images. Like Ian Goodfellow's GANs, Tenenbaum's method employs another level of feedback, in his case the connection between motor efference and visual perception. Adding layers of feedback seems to improve the performance of at least some AI tasks.

Today, computers have limited abilities to "do" much in the physical world. Very few computers are equipped with arms and hands that can pick up a pencil and draw a character on paper (for an exception, see the Turing machine in figure 8.1, which mechanically draws the characters 0 and 1 on a strip of film leader). But with the Internet of Things (IoT) revolution, this situation is changing quickly. The IoT connects many physical sensors and actuators to the world of computers and networks, endowing the computers with eyes, ears, hands, and feet. If these sensors and actuators are recruited to enable more physically embodied learning-by-doing, I suspect the advances will be dramatic.

With more layers of feedback, actuators that affect the physical world, and perhaps even analog components, technology will inevitably acquire greater cognitive capabilities. Will it reach the point where we have to hold machines accountable for their actions? This is the topic of the next chapter.

10

ACCOUNTABILITY

WHO IS THE ARTIST?

On October 25, 2018, the art auction house Christies put on sale a painting called *Portrait of Edmond de Belamy* that was expected to fetch between $7,000 and $10,000 (see figure 10.1). The painting was created by an artificial intelligence (AI) using a variant of Ian Goodfellow's GAN algorithm (see chapter 4). Surprising everyone, the painting sold for $432,000.

The code to create the painting was written by Robbie Barrat, who started the project when he was a seventeen-year-old West Virginia high school student. He subsequently went to Stanford to work on AI. Barrat has used his code to generate beautiful and surreal landscapes and nudes, including some highly dynamic artworks that morph from one image to another in ways that no traditional painting can do.

Barrat put his code on GitHub, a website for collaborative code development, where it was picked up by three French twenty-five-year-olds who run an art collective that they call OBVIOUS.[1] Pierre Fautrel, Hugo Caselles-Dupré, and Gauthier Vernier say that the mission of their collective is to "explain and democratize" AI and machine learning through their artworks. They made slight modifications to Barrat's code and fed the neural net some fifteen thousand portraits scraped from the web dating from the fourteenth through the twentieth centuries. They then let the algorithm run and selected among the results.

10.1 Pierre Fautrel of OBVIOUS standing next to a work of art created by an AI algorithm entitled *Portrait of Edmond de Belamy.* The portrait sold at Christies in New York for $432,000. Photo by TIMOTHY A. CLARY/AFP/Getty Images.

So who is the artist? According to the online journal *The Verge*, there has been some controversy and no shortage of jealousy and hurt feelings:

Mario Klingemann, a German artist who has won awards for his own work with GANs, tells *The Verge* over email, "You could argue that probably ninety percent of the actual 'work' was done by [Barrat]." Tom White, an academic and AI artist from New Zealand, says the work is extremely similar, even downloading Barrat's code and running it with zero adjustments to compare the outputs.[2]

Part of the reason that OBVIOUS has received more credit for this work than Barrat, according to *The Verge*, is their narrative crediting the AI for the work, saying "creativity isn't only for humans." Such a conceptual art position reflects well-established practice in the art world, where an artist can, for example, declare a found object to be a work of art. The most notable examples are Marcel Duchamp's readymades. If Duchamp declares a bicycle wheel to be a work of art, who really is the artist?

Computer-generated art is a bit like a found object. Any given program will produce a vast number of candidate outputs, and selecting among them is part of the creative process. Duchamp too had many choices of found objects. Moreover, selecting from among choices over which the artist has little control is not much different from a painter, upon dripping or splattering paint onto a canvas, choosing to accept or reject the outcome. Or perhaps we should view Barrat more like a paint maker than an artist, a craftsman who made the medium, not the artwork. To be fair to Barrat, however, his contribution is much more than to just provide the medium, since his own selections of AI-generated art are spectacular. He provided not only the medium, but also the inspiration. But once again, emulating another inspiring artist is a well-established practice in the art world.

Even with the work of *humans*, assigning accountability is far from trivial. It seems to me that I chose to write the words on this page, and that you can and should hold me accountable for them. But could I really have chosen other words? Perhaps these words are simple consequences of the state of my brain and the room I am sitting in. The sun through the window is warming my head, and perhaps that is why these words are coming out as they are. How can we check that hypothesis? Is there an experiment that will determine whether I chose these words over others? If I close the blinds, will different words come out?

If my computer loses these words, something that does occasionally happen, then I will be unable to recreate them exactly. The thoughts I am attempting to convey with these words, contrary to Nietzsche (see chapter 4), are not in fact isomorphic with these words. If the words are lost, have the thoughts been lost too? I can go back and edit the words I wrote, but that does not change the fact that I wrote what I wrote even if all history of that fact has vanished. If the words are changed, have the thoughts also changed?

Could a computer have written these words? If you, dear reader, are an AI, could you have written these words? More fundamentally, could you have *chosen* to write these words over alternative words with a different message? If an AI had written these words, would you hold the AI accountable for them? Or the programmers of the AI? Or the computer hardware that runs the program? Or the electric company that feeds the hardware? Can we hold an inanimate process accountable for anything? Is an AI inanimate?

CRASHES AND VIRUSES

Figure 10.2 shows the aftermath of a rear-end collision, where a car has run into the back of an SUV. The car, apparently, took no action to prevent the collision. Should we hold the car accountable? Is inaction as culpable as action? With today's technology, there really is no excuse for putting cars on the road that will do this. The technology to prevent collisions of this type is readily available, inexpensive, and reasonably reliable.[3] Today, we do not hold cars nor their manufacturers accountable for inaction, but maybe we should. That car should be ashamed of itself. Car manufacturers today even put cars on the road that will take no action to prevent such a collision even though they already have all the hardware onboard that would be needed to prevent the collision, such as a forward-looking radar used by the adaptive cruise control system.

The classic trolley problem poses the question of the ethics of inaction. The trolley problem is a thought experiment where you are faced with a situation where inaction leads to a bad outcome and action leads to a less bad, but still bad outcome. In the usual formulation, you see a runaway trolley moving toward five people lying on the tracks. You are standing next to a lever that controls a switch. If you pull the lever, the trolley will

10.2 The grey car should be ashamed of itself.

be redirected onto a side track with a single person lying on it. If you do nothing, five people die, and if you pull the lever, one person dies. But if you pull the lever, it will have been your action that caused that one person to die. Which is the more ethical option?

Once, after a talk that I gave, I got a question from someone in the audience who asked how we can ethically deploy self-driving cars without first solving the trolley problem. How should the car react, for example, if staying on its current course will kill the passenger in the car, but swerving to avoid the collision will kill a pedestrian. My answer was another question: how can we deploy human-driven cars without first solving the trolley problem? We haven't solved this problem for humans, so it's hard to see how having human drivers is any better than having computer drivers, at least with respect to this problem. The reality is that humans and machines alike will face scenarios that have no good outcomes, and no amount of advance planning is going to ensure that the least-bad outcome is chosen every time. Not having solved these problems is no excuse for delaying deployment of technology that could save thousands of lives.

An online study carried out by a group of MIT researchers in 2018 showed that humans have significantly different moral preferences when faced with having to choose between two bad outcomes.[4] Some of these preferences appear to be culturally determined, for example on the question of whether to spare the life of a child or an older person. In the study, however, humans had plenty of time to ponder the question of how they would prefer to act in such a situation. Such situations in real life usually happen so quickly that no human could possibly rationally evaluate the alternatives. A computer, however, can evaluate alternatives far faster, making the trolley problem and problems like it more relevant for self-driving cars than for humans. With self-driving cars, we could possibly ensure that *someone's* moral preferences are applied, even if not everyone would agree on the choice. With humans, given our limited reaction time, we really can't even ensure that the driver's own moral preferences would be applied.

Simpler ethical problems than the trolley problem also might be more easily solved in software than in humans. Yuval Noah Harari, in his most recent book, *21 Lessons for the 21st Century*, points out that otherwise ethical humans often behave unethically due to anger, lust, or even just the pressures of daily life, whereas computers can be programmed to reliably follow rules encoding ethical behavior.[5] This may be harder than it looks, however, because software will be increasingly built on inscrutable machine learning algorithms rather than the rule-following GOFAI of the 1980s (see chapters 4 and 6). Moreover, reducing ethics to unambiguous imperative rules is not easy.

Intellects far greater than mine, over centuries, have addressed, though not resolved, astonishingly difficult questions of ethics and accountability. It would be hubris to believe that I have anything to add. With few exceptions, though, these questions have had a human face. Those higher intellects discuss whether *humans* have free will and can be held accountable for their choices.

The Berkeley philosopher John Searle has dubbed the free will problem "a scandal in philosophy" on which we have made little progress since antiquity.[6] Today, we confront, without metaphor or allegory, the question of whether a machine, specifically software executing on a digital machine, can be held accountable for its actions. Maybe the problem is easier to solve, or at least to understand, with machines rather than humans.

As a culture, we have a strong need for accountability. When something good or bad happens, we need someone or something to credit or blame. When we can't find one, we may invent a God to hold accountable. Chance, a devil lurking in the shadows, is a last resort, getting the credit or blame when nobody else can be found.

On October 19, 1987, now known as Black Monday, the Dow Jones Industrial Average fell 22.6 percent, the largest one-day drop in history, another kind of crash. Although many factors contributed to this global stock market collapse, many economists and pundits blamed computerized trading, and particularly an algorithmic strategy known as portfolio insurance.[7] Portfolio insurance, ironically, was designed to protect investors against substantial market drops, but it is clear that in this case, the algorithm itself at least contributed to the drop, even if it didn't actually cause it.

In 1987, automated trading was far less prevalent than it is today, and it is probable that much of the price collapse was driven by humans, not by computers. On May 6, 2010, however, the Dow dropped nearly one thousand points (nine percent) and then recovered, all within about

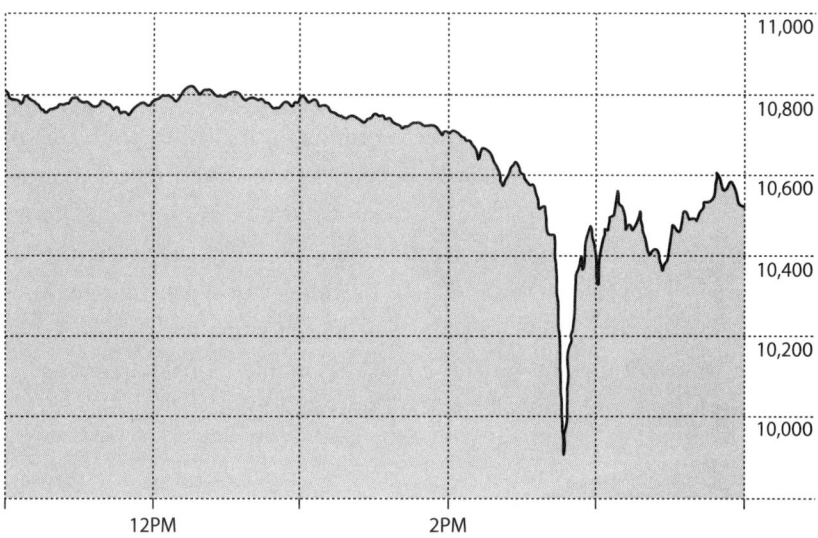

10.3 The flash crash of the Dow Jones Industrial Average on May 6, 2010.

thirty minutes (see figure 10.3). Known as the flash crash, this event was much too quick to be mainly due to human-driven trades. After five years of investigation, the US Department of Justice indicted Navinder Singh Sarao, a trader working from his parents' house in suburban west London, for fraud and market manipulation. He had allegedly modified commercially available software to perform rapid placement, modification, and cancellation of options trades. Although most observers do not blame Sarao exclusively for the crash, it is likely that his software contributed to it, and hence he provided a convenient focus point for accountability. But what if the software had been modified by other software instead of by humans? Where then do we put the blame? That is the question we are facing today.

In chapter 2, we met Richard Skrenta, who created Elk Cloner, the first virus that successfully targeted personal computers, when he was in the ninth grade. Skrenta had his fifteen minutes of fame for this, so he was certainly held accountable, but he did not suffer any adverse consequences for what he ultimately viewed as a practical joke. In notable contrast, Jeffrey Lee Parson, an eighteen year old, got an eighteen-month prison sentence for the Blaster computer virus, which he modified but did not author. The original author was never found. Humans are not very consistent about holding other humans accountable. Can we expect any better from machines?

Today, we use AIs to detect malicious software, but what's to say that they would not be just as good at synthesizing malicious software? In chapter 4, we encountered Google's DeepDream, which reversed an AI classifier trained to recognize dogs to synthesize bizarre mutant images (see figure 4.3). Why not reverse the malware classifiers to synthesize computer viruses? DeepDream may be harmless, but synthesizing viruses would not be. It is quite possible, I believe, for malware synthesis to emerge in a way that no human could reasonably be held accountable. Please indulge me, dear reader, while I spin that tale.

A TALE OF TANGLED ACCOUNTABILITY

Most computer programs have more than one author. Many programs have many authors. Having many authors dilutes accountability. If the number of authors gets large enough, identifying those to hold accountable will become more difficult.

Many programmers collaborate on programs using GitHub, a web-based service that was acquired by Microsoft for 1.7 billion US dollars in 2018. Many programmers can work on the same program simultaneously thanks to Git, the heart of GitHub.[8] A computer program is a text file, not unlike the file storing this book on my computer. The text in the file is the "DNA" of the program, the instructions defining how the machine works when it comes to life. The programmers are the forces of mutation for this genetic material.

In a collaborative software development project, several people (and maybe even a few AIs) may be simultaneously working on the same program, changing its genes. It can become easy to for these programmers to step on each other's toes. Version control systems such as Git help to prevent this by maintaining a common, shared version of the program and merging the changes from multiple programmers.

One of the most interesting features of Git is that it permanently stores all historical variants of a program as the program evolves. Moreover, for each change that is made to the program, it records who made the change, or more precisely, which GitHub account was responsible for the change. Imagine if biological evolution had used GitHub and we had a record of every mutation ever made to our DNA, including an accounting of which virus or radioactive source caused the mutation. That record would be quite a bonanza for evolutionary biologists! For many computer programs, that record exists for anyone to read. In principle, therefore, it should be easy to assign responsibility for any change that causes the synthesis of malware. But is it really so easy?

In chapter 5, I speculated that machines would acquire ever deeper layering of feedback loops. Programs that learn to classify images, for example, are such feedback loops, and they can be turned into programs that synthesize, as done by Google with DeepDream. We could layer these loops and create an AI that recognizes programs that learn to classify.[9] Let's name this meta classifier "DeepClassifier." Given the source code for a program as input, DeepClassifier will tell us whether it is a program that uses deep learning to classify its inputs into some set of categories. For example, given the source code for Google's Inception, which we encountered in chapter 6, DeepClassifier would tell us, "yes, it's a classifier using deep learning."

Once we have created DeepClassifier, we could unleash it on GitHub. As of 2018, GitHub had twenty-eight million publicly readable repositories, most of which contain computer programs. Many of these programs are deep classifiers, though honestly I have no idea how many. DeepClassifier will be able to answer that question and create a collection of such programs.

Once we have DeepClassifier, there are many potentially interesting uses for it. One use would be to mutate each of the collected classifiers, turning it into a synthesizer like DeepDream. This mutation could probably be automated in a program that we might call DeepMutator. DeepMutator may have entirely well-intentioned purposes. For example, such a program could be used to create ever richer training data sets for the programs that learn to classify. Or it could be part of strategy to create an "explanation" for the classifications that an AI comes up with, solving DARPA's Explainable AI challenge (see chapter 6). Or it could be a way to improve the generative adversarial networks that we saw in chapter 4. Or it could help to create tests for deep learning programs.

A combination of DeepClassifier, a tool to scour GitHub repositories, and DeepMutator will likely stumble across a classifier that classifies malware and turn it into a synthesizer that synthesizes malware. Who will be held responsible when the resulting malware infects millions of machines worldwide? Finding a programmer to blame for this outcome may become analogous to finding the neuron in a murderer's brain to blame for the murder. Even with all the traceability provided by Git, it will be like all the king's men trying to put Humpty Dumpty together again.

If such an event occurs, we are unlikely to be able to pin accountability on any individuals or corporations. We will have to treat the event as an "act of God" or, more rationally, as an epidemic, where many of the machines that we depend on in our symbiotic ecosystem have been infected by a disease whose root cause is another species of machine that evolved in a Darwinian way. Very likely, the infected machines' immune systems, if they even have immune systems, will be inadequate, and human intervention will be required to eradicate the disease. For such intervention, the detailed record in GitHub could prove valuable by helping us find a cure, even if it doesn't help us find a culprit. We humans could midwife new machines that will scour the GitHub record to synthesize effective countermeasures, which could even take the form of "genetically" engineered machines that spread themselves in an epidemic-like fashion, kill the pathological machines, and then commit suicide. None of this, however, will lead us to a single culprit, nor even a scapegoat.

VOLITION

One way to approach the problem of accountability is to associate accountability with volition, the power to freely use one's will. We hold an individual responsible for an outcome if the individual chooses an action that leads to that outcome. A common legal test of a defendant's culpability is called "but-for causation"; the injury would not have occurred but for the defendant's action. The defendant is not culpable without choice, however. Consider a person that is pushed out a window and falls on a passerby. Falling is arguably an "action," but we would not hold that person responsible for the injuries of the passerby, even though there would be no injuries but for that action. But-for causation, therefore, is better read as the injury would not have occurred but for the defendant's *chosen* action. If machines are not capable of choice, then by this criterion, they can never be culpable.

We still need to make a distinction between a choice that is *intended* to produce that outcome and one that *accidentally* leads to that outcome. A person who jumps out a window and accidentally lands on a passerby may be held accountable for injuries. The legal system in most countries makes a distinction between manslaughter and murder, for example, and between voluntary and involuntary manslaughter. Even involuntary manslaughter, however, requires choice. A person chooses to drink, then chooses to drive, and then runs off the road and kills someone. We still require choice to hold the person responsible for involuntary manslaughter. Are machines capable of choice? Are they capable of intention? How can we know?

Philosophers have long struggled with the question of whether humans are capable of choice. Is this famous "free will" question easier to answer for machines? A glib answer is that software is an automaton or a deterministic process and therefore cannot have free will. But what if the world is deterministic, and humans are just automata too? Our actions are then dictated by the laws of physics and biology, not by any volition. Should we hold those laws accountable instead of holding humans accountable?

Here, some care is required when talking of determinism. In physics, a nondeterministic event is one that has no cause. (See chapter 11 for the notion of causes.) Its outcome cannot be explained by antecedent events or states. In contrast, in computer science, a nondeterministic event is one that has more than one possible outcome, regardless of what causes one outcome to be selected over another. In computer science, therefore, an event may be nondeterministic simply because we do not know what causes it, whereas in physics, it is nondeterministic if there is no cause. It just happens. These two uses of the same term can lead to confusion. I will try to be clear which way I'm using these terms each time, but if I do not say, please assume I am using the physics sense.

Daniel Dennett, one of the many intellects far greater than mine who have addressed the question of free will, argues convincingly that even if the world is deterministic (every event follows from a cause), free will exists and humans can and should be held accountable for their actions.

Does this imply that AIs must also be held accountable? Most AI programs are deterministic in this sense. Their behavior is a consequence of the way they are programmed and the inputs provided to them. Dennett has squared off against Sam Harris, a very interesting polymath with a PhD in neuroscience and a successful career as a writer and philosopher. Harris describes himself as "someone who thinks in public," and he fearlessly and publicly confronts controversial issues with a powerful intellect.

Along with Dennett, Richard Dawkins, and Christopher Hitchens, Harris is described as one of the "Four Horsemen of Atheism." His 2004 book, *The End of Faith: Religion, Terror, and the Future of Reason*, was motivated by the events of September 11, 2001, and lays the responsibility for those events at the feet of organized religion. Notwithstanding his atheism, Harris has a deeply held moral compass and no shortage of spirituality of a sort.[10] He dropped out of Stanford and went to India and Nepal to study meditation with Buddhist and Hindu religious teachers. He later returned to Stanford to major in philosophy, and then went to UCLA to get a PhD in neuroscience. In 2018, among his many activities, he released a meditation app for smartphones. Meditation is fundamentally a first-person experience; it requires engagement of the self and cannot be meaningfully externally observed. In the following two chapters, I will use the ideas of no fewer than four Turing Award winners to show why first-person experiences are able to accomplish things that no third-person observation can. Are computers capable of first-person experience, and is that a prerequisite for holding them accountable for their actions?

Sam Harris's politics are also complex. He debates right-wing pundits, takes decidedly liberal positions in his podcast, and yet blogs in favor of gun rights, a position commonly associated with the right. He admits a Twitter addiction, which in Fall 2018 he said he was attempting to curtail, and he has received numerous death threats for his views on politics and religion. I have immense admiration for his intellect and his courage, so it is with some trepidation that I find myself disagreeing with his position on free will. Fortunately, I believe that Daniel Dennett, no less a great intellect, is on my side. But anyway, my intent here is not to stick my nose into this debate, but rather to see how it can help us understand whether and how to hold machines accountable for their actions.

In his 2012 book *Free Will*, Sam Harris ties the notion of free will tightly to accountability:

Without free will, sinners and criminals would be nothing more than poorly calibrated clockwork, and any conception of justice that emphasized punishing them (rather than deterring, rehabilitating, or merely containing them) would appear utterly incongruous.[11]

So far, I completely agree. But he goes further to assert definitively that the notion of free will is dead:

Free will is an illusion. Our wills are simply not of our own making. Thoughts and intentions emerge from background causes of which we are unaware and over which we exert no conscious control. We do not have the freedom we think we have.

Free will is actually more than an illusion (or less), in that it cannot be made conceptually coherent. Either our wills are determined by prior causes and we are not responsible for them, or they are the product of chance and we are not responsible for them.[12]

He goes on,

We are not the authors of our thoughts and actions in the way that people generally suppose.[13]

But who (or what) is this "we" that Harris says cannot be responsible? Here, Harris is talking about a commonsense notion of "I," my conscious mind, and he relies on experiments from psychology and neuroscience that show that the conscious mind does not originate decisions:

The physiologist Benjamin Libet famously used EEG [electroencephalogram] to show that activity in the brain's motor cortex can be detected some three hundred milliseconds before a person feels that he has decided to move.[14] Another lab extended this work using functional magnetic resonance imaging (fMRI): Subjects were asked to press one of two buttons while watching a "clock" composed of a random sequence of letters appearing on a screen. They reported which letter was visible at the moment they decided to press one button or the other. The experimenters found two brain regions that contained information about which button subjects would press a full seven to ten seconds before the decision was consciously made.[15] More recently, direct recordings from the cortex showed that the activity of merely 256 neurons was sufficient to predict with eighty percent accuracy a person's decision to move seven hundred milliseconds before he became aware of it.[16]

These findings are difficult to reconcile with the sense that we are the conscious authors of our actions. One fact now seems indisputable: Some moments before you are aware of what you will do next—a time in which you subjectively appear to have complete freedom to behave however you please—your brain has already determined what you will do. You then become conscious of this "decision" and believe that you are in the process of making it.[17]

These studies do indeed show definitively that the conscious mind is not a conscious homunculus driving the bus, so to speak. But what if consciousness itself is a byproduct of the mind, emerging delayed from the neurobiological processes that constitute the real "we"?

Applying these insights to computers and software is difficult because it would require them to have a "conscious mind" that we can then assert is not in control of their decision making. I do not believe that any AI today has anything like a conscious mind, but it is not obvious that they won't in the future. Unfortunately, as I will argue, if they ever do acquire something resembling consciousness, we probably will not and cannot know it. Actually, I cannot really even be sure that any human other than myself has consciousness either. I have no direct evidence. I *infer* that humans are conscious because they resemble me. But computers are not similar to me and probably never will be, so it will be much more difficult to come to the conclusion that they too have developed a conscious mind. It is not clear, therefore, that Harris's reasoning about free will can help at all in guiding decisions about how and whether to hold machines accountable for their actions.

IMAGINING ALTERNATIVES

Arguably, you can only hold me accountable for the words on this page if I could have chosen different words with a different message. I believe that I could have, but is this belief well grounded in the facts of the physical world? "Belief," after all, is a learned mental state, and this belief could arise from my sense of agency, which in turn could arise from my development as an automaton that functions effectively in the physical world. My notion of causation, that my "I" can cause the words on this page, may be an illusion, a sensation that my nervous system has created because, without that sensation, my species would have become extinct. Is Sam Harris right? He says that if the world is deterministic, there is no free will, and if the world is nondeterministic, there is still no free will.

Or is Dennett right, that my belief in my free will is what we ultimately mean by "free will," and since I believe I have free will, it is OK with me if you hold me accountable?

For these questions to make any sense, we have to at least *imagine* that things could have been different, that there are alternatives, that I could have written different words. Fortunately, our brains are equipped with the basic mechanism needed for such imagining, a mechanism that we can call a "fake efference." I *imagine* words that could be on the page, sound them out in my head without making any physical sound, hear how graceful or how awkward they sound, and accept or reject them. The "sounding out in my head" is a fake efference; it does not actually cause my larynx to do anything to produce sound. The voice I hear in my head is a fake reafference,[18] not a signal picked up by my ears, but rather synthesized by my brain and fed back into the part of my brain where it would have appeared had it been a signal picked up by my ears. The language centers in my brain process that fake reafference and form a judgment about the elegance, flow, and understandability of the words. If I like the result, my brain issues the motor efferences for my fingers to type the words. The expected reafference pops up physically in front of me as words that I see on the screen. I then subject those words to further scrutiny, converting the visual stimulus into language and again sounding them out in my head. I then revise, for the twentieth time, this paragraph.

My brain is equipped to cogitate on many alternative word sequences before I take any physical action. But even that cogitating is composed of physical actions, and how alternatives manifest themselves in the brain is a bit of a mystery. Just now, for the previous sentence, my brain somehow came up with "manifest themselves" and "appear" as alternative wordings. I chose "manifest themselves" over "appear." How did these two alternatives appear (manifest themselves?) and why did I pick the one I picked?

There are two opportunities here for chance to play a role. The first is that it may have been chance that made the two alternatives, "manifest themselves" and "appear," appear. Many other alternatives may have manifested themselves, such as "crop up," "develop," "emerge," "materialize," "occur," "pop up," "present themselves," "show up," "surface," "turn up," "spring forth," and "arise." None of those alternatives popped into my head. I had to use a thesaurus just now to come up with this list.

Once the two alternatives, "manifest themselves" and "appear," sprang forth, a second opportunity for chance to play a role materialized. I could have selected "manifest themselves" by random chance, perhaps as a result of a quantum wave function collapsing somewhere in my brain. More likely, I think, these two alternatives fed back as fake reafferences, sounded out in my head, and my brain liked "manifest themselves" better than "appear." Perhaps the pleasure centers in my brain were more strongly stimulated by "manifest themselves" than by "appear."

If the first development was by chance, Sam Harris would say that there was no free choice because my "I," my conscious mind, played no role in making these two particular alternatives present themselves. If it was not by chance, but preordained by the state of the universe and the laws of physics, then again there was no free will involved. Daniel Dennett would say that whatever mechanisms in my brain caused these two alternatives to crop up, chance or brain physics, it was still my brain, an essential part of my "I," that made these particular two alternatives surface, so it was my "I" that resolved that these were the two possibilities over many others.

Whether the second stage, the selection of "manifest themselves" over "appear," was by chance or preordained, Harris would argue that the selection did not involve free will. I suspect, however, that this choice was not entirely chance. The pleasure centers of my brain were probably involved, at least to some degree. Those pleasure centers, and how they react to words, are an essential part of what "I" is, having been finely tuned by expensive education to prefer elaborate verbiage over simpler words. Dennett would say that the fact that "I" resonated more strongly with "manifest themselves" than with "appear" is the essence of what we mean by "free will," even if that resonance was already wired into my brain, even if the choice was the result of deterministic physics working its way through that brain, and even if the choice was made before my conscious mind became aware of it. That same brain (mine) sides with Dennett over Harris on this issue. I did freely choose the words "manifest themselves," even if that choice was a result of my expensive education. That my conscious mind became aware of the decision only after it was made is not so important to me because my conscious mind is just another manifestation of the biophysics that is me, a part of a bigger picture.

When a deep neural net synthesizes a sentence, for example, when Alexa "chooses" a response, a similarly nuanced process occurs. The numerical weights in a neural net, learned via backpropagation over millions of examples, end up giving slightly higher scores to one candidate response over another. Those weights constitute the "I" that is Alexa at that moment in time, giving her her "personality" and reflecting her "education."

WHEN, WHETHER, WHY, AND HOW

The two alternatives, "manifest themselves" and "appear," at some point in time, became the two wordings for me to consider, somehow. If the world is deterministic (in the physics sense), then these two particular words were direct consequences of the prior state of the universe. That prior state of the universe was a direct consequence of the state before that, and before that, all the way back to the Big Bang.

If on the other hand the world is nondeterministic, the two wordings popped into existence in my brain just shortly before I became conscious of them as a result of some uncaused event, perhaps a quantum phenomenon, but also possibly a simple classical metastable state resolving itself (I will discuss metastable states in chapter 12). Then I chose "manifest themselves" over "appear," a choice that may have involved chance, deterministic biophysics, or some combination of the two. In both cases, whether the world is deterministic or not, there were alternatives. In one case, the resolution of alternatives was (perhaps) made at the time of the Big Bang, and in the other case, the resolution was made much later.[19] Determinism (in the physics sense), therefore, is about *when* the resolution of alternatives occurs, not about *whether there are alternatives*. In a deterministic universe, the choice between alternatives is *resolved* much earlier, perhaps at the time of the Big Bang. In a nondeterministic universe, resolution occurs much later.

The question of free will, however, is more about *how* and *why* the resolution of alternatives is made. Harris would argue that if the resolution occurs at the Big Bang, then human consciousness played no role in the resolution, so humans cannot have free will, at least not in the sense of conscious decision making. Dennett would argue that even if the resolution was predetermined at the time of the Big Bang, the biophysical processes that carried out the manifestation of the resolution occurred much later, and those biophysical processes are what we mean by "I," so it is "I" who made the choices.[20] The biophysical processes that are "I" made me use the words "manifest themselves" in this book.

Daniel Dennett goes to great lengths to show that determinism is not incompatible with free will.[21] Such so-called "compatibilism" has been argued since at least the ancient stoics, the medieval scholastic Thomas Aquinas, and the enlightenment philosophers David Hume and Thomas Hobbes. Sam Harris, on the other hand, is a contemporary incompatibilist; he argues forcefully that free will is incompatible with *both* a deterministic and a nondeterministic world.[22]

I will return to the question of determinism from the perspective of physics later, where I show that our best physical theories today do in fact admit nondeterminism even if many physical processes are best modeled as deterministic. Computers, however, operate within a subset of physics, the algorithmic and digital world that is almost always deterministic. Turing machines are deterministic, and any nondeterminism that arises in any physical realization of a Turing machine has to be considered a fault condition, a failure of the Turing machine, an error.[23] If such an error occurs, it occurs "outside the system," not within the algorithmic and digital world. Therefore, when deciding whether machines can have free will, if we restrict our attention to those machines made from Turing machines or their equivalent, then they can only have free will if the compatibilists, like Dennett, are right.

Harris, on the other hand, appears to assert not only that humans do not have free will, but that there cannot possibly be any mechanism at all in the physical world having free will, in the sense of conscious decision making. If the world is deterministic, "no choices" implies no free will, according to his incompatibilist perspective. If choices are resolved by chance, on the other hand, not driven by any decision maker or other cause, then again there is no free will. In both cases, the conscious mind learns about the decisions only after they have been made, and therefore cannot be in control. For there to be free will, a choice needs to be resolved (caused) by a decision maker. If that decision maker is a mechanism, it cannot be a deterministic mechanism, or we are back with incompatibilism. But it also cannot use nondeterminism to make the choice because that would also undermine its free will. Unless we reject the law of the excluded middle, there cannot be any middle ground between determinism and nondeterminism.[24] In the words of Yuval Noah Harari,

To the best of our scientific understanding, determinism and randomness have divided the entire cake between them, leaving not even a crumb for "freedom."[25]

As a consequence, in effect, Harris tells us that *no* mechanism can have free will. Thus, today's machines do not have free will, and no machine invented in the future will have free will. Free will is simply something that does not and cannot exist in the universe, according to Harris.

If Harris is right, then we should change the question. In this case, the question we should ask is not "can machines acquire free will?" It is instead "can machines acquire sufficient agency that we should hold them accountable for their actions?" To me, this is just more verbiage for the same question, so I find Harris's position not very useful.

If we accept Daniel Dennett's position, you can hold the biophysical processes that constitute "me" responsible for the words on this page, regardless of whether the world is deterministic. Unless you believe that free will can only be biological, this may lead to a conclusion that you would have to hold any machine responsible for any words or actions it creates. But this gives too simple an answer. I suspect that Dennett would object to this answer on the grounds that his interpretation relies on a "common sense" notion of what we mean by "free will." If that common sense notion is inextricably tied to biology, then free will is not a property applicable to machines. We are again forced to change the question in the same way.

The only way out of this conundrum that I can find is to simply assert that henceforth in this book, when I use the term "free will," I will mean "sufficient agency that we should hold them accountable for their actions." But what could we possibly mean by "sufficient agency"?

VULGARITY AND RACISM

On March 23, 2016, Microsoft released a chatbot called Tay, an AI designed to interact with users on social media such as Twitter using the vernacular of hip youngsters using the media.[26] Tay was designed to learn on the fly, and unfortunately, she did that very well. Tay was quickly trained by (probably malicious) humans, through Twitter, to write vulgar and racist tweets. According to James Vincent,[27] in an emailed statement given to Business Insider, Microsoft said, "The AI chatbot Tay is a machine learning project, designed for human engagement. As it learns, some of its responses are inappropriate and indicative of the types of interactions some people are having with it. We're making some adjustments to Tay." Microsoft euthanized Tay.

It is true that humans played with Tay in a nasty way knowing that she was a robot. Perhaps Google's Duplex (see chapter 9) can be similarly trained to be nasty, and when two Dupes are talking to each other, they will just reinforce each other's nastiness in a runaway feedback loop of supervulgarity rather than superintelligence. Will humans similarly play with self-driving cars, cutting them off on the freeway or stepping out in front of them to watch them slam on the brakes? There are documented studies that show that humans sometimes behave abusively toward robots in ways that they would not treat other human beings.[28]

If Tay can be held accountable for her vulgarity and racism, then perhaps Microsoft was justified in executing her. Or is "executing" too strong a word? If Tay was living while tweeting, then shutting her down certainly seems like an execution, although I'm sure Microsoft could resurrect her from backups. But even so, maybe it is no more an execution than what we do when we take antibiotics, killing bacteria. Or is it "killing" at all? We are entering an era where we cannot continue to ignore these questions.

Perhaps we should hold Tay's programmers responsible for her words, but I suspect they were no more vulgar and racist than the average person. The programmers were likely just as surprised by the words as we were. Can you be held accountable for the actions of your kids? You created them and participated intensively in their formation as human beings, and in many cases, humans have decided that yes, you can be held accountable. But in other cases not. Navinder Sarao was probably surprised by the flash crash that emerged while his software was running, but the Department of Justice did not think that absolved him of responsibility. And his parents were not held responsible, even though he was operating out of their house.

In *Timaeus*, one of his most influential dialogues, Plato argues that "diseases of the soul" have a physical rather than divine origin. Wrongful acts are caused by defective bodily constitution or faulty education. In Plato's reasoning, wrongful acts should not lead to reproach because the acts are not willful. In other words, if the cause for the act is physical, the actor should not be held accountable. If we *understand* what causes the act (brain damage, mental disease, poor upbringing, and so on), then the actor is less culpable than if we do not understand what causes the act. I believe that this philosophy still dominates today.

Tay used deep learning. We humans have a pretty good understanding of the mechanisms of machine learning, but as discussed in chapter 6, we often cannot provide satisfactory explanations for the behavior of these algorithms. Since we understand the underlying mechanisms, however, it is clear that Tay's racist and vulgar tweets cropped up due to physical causes, not divine ones. Therefore, by Plato's logic, Tay should not be held accountable.

This conclusion, however, inevitably takes us to a place we probably do not want to be. If cognition is a material phenomenon, due to biology and physics, as I believe it is, then everything we humans do has a physical origin and we cannot be reproached for our actions. In this view, it does not matter whether the origins are deterministic or nondeterministic. It only matters whether we understand what the origins are. For now, we are saved by the fact that we do not understand the biology and physics of the brain well enough to be sure how it works, unlike Tay's situation. But does this mean that as our understanding of neuroscience improves, free will and accountability both evaporate?

MACHINE CREATIVITY

In addition to his argument that the conscious mind is not in charge of decisions, Sam Harris states that if choices are made at random, then free will is not involved:

If my decision to have a second cup of coffee this morning was due to a random release of neurotransmitters, how could the indeterminacy of the initiating event count as the free exercise of my will? Chance occurrences are by definition ones for which I can claim no responsibility. And if certain of my behaviors are truly the result of chance, they should be surprising even to me. How would neurological ambushes of this kind make me free?[29]

At a certain level, the answers to Harris's questions are obvious, but what if choices are made in a more complicated way? The two alternatives, "manifest themselves" and "appear," may have shown up at random, but the selection between them may have involved my expensive education, resonating with a brain whose structure and mechanisms have been built up over many years to make what is undeniably me. Harris's vision, I believe, is that if the world is nondeterministic, then I flipped a coin, one side of which said "manifest themselves" and the other "appear." Such a one-step "decision" is clearly not a decision and does not involve free will.

Decision making in humans, however, is much more complicated. We envision alternatives, creating fake reafferances and weighing how they affect our brain. New alternatives almost certainly crop up at random, but then they get judged by machines, our brains, whose preferences are personal. There are many more steps than Harris's random release of neurotransmitters implies.

Every artist knows that chance accidents can drive creativity. But these accidents are not, as Harris assumes, discrete, monolithic, one-step events. It is not that Picasso's brain, buffeted by quantum randomness, suddenly envisioned *Les Demoiselles d'Avignon*. Boom. Out pops a whole painting. No, it's more like the paint dripped here and mixed there, and with each flex of the muscles in his hand, Picasso saw something unexpected and delightful. Small continual accidents please the artist and help to shape the work, whereas big ones, ones that Harris would not associate with free will, are more likely to result in throwing the painting away as ruined. The continual feedback between the artist's brain, with its own internal random nudges, and the physics of paint and canvas, with its external random nudges, combine to create the feeling of control, the personal ownership of the work, despite the randomness. As part of the feedback, the artist's brain accepts, rejects, or, more commonly, tweaks the continual accidents, imparting on the painting a personal style, one governed by what delights this particular brain. That "delight" is so much dopamine, influenced by that brain's own particular character, which has been shaped by genes and years of experience. It is easy to see that such randomness plus feedback can form the core of what we call "creativity."

Creative people tend to be risk takers more than usual. Perhaps they have brains that are more open to randomness. Perhaps their brains use a higher gain in the feedback loop, sometimes overcorrecting, sometimes disrupting the physical world into a new and completely unexpected mode. To be creative, that brain must recognize the benefits of the newly disrupted system. It must derive pleasure from the disruption in order to stay in that unexpected mode, to stop or slow the feedback corrections to the controller that would bring the whole interaction back into the realm of the expected. Picasso had to see the drips of paint as fortuitous, not annoying.

Today, computers have some handicaps compared to humans when it comes to creativity. First, their designers have worked very hard to keep randomness out. Turing machines are deterministic, and when randomness creeps in, due for example to concurrency, it is usually considered a bug. The algorithm used to make the *Portrait of Edmond de Belamy* probably had no randomness in it. Nevertheless, randomness probably played a role in the brains of Fautrel, Caselles-Dupré, and Vernier when selecting among the thousands of outputs from the program the one image they would declare to be their conceptual art creation. The collection of images fed into the algorithm, harvested from the Internet, probably also was effectively random.

A few programming styles explicitly embrace randomness. So-called randomized algorithms, for example, inject into an otherwise deterministic program random choices. This can help an algorithm to explore a large space of possibilities without getting stuck in a small local cluster. One project explicitly embraces randomness to achieve what I am tempted to call "machine creativity," but which the inventors call "control improvisation."[30] Sanjit Seshia, a colleague and friend at Berkeley, has been leading a collaborative project in which I am lucky to play a small role. The first application was to jazz improvisation.[31] The algorithm would learn a musical style by analyzing examples, then inject randomness into synthesized variants, constraining the synthesized sounds with models that kept the pitches and rhythms reasonably close to a reference melody. Seshia and his students have extended this technique to get computers to exhibit human-like behavior in a wide variety of scenarios, including, for example, automated driving and controlling the lights in a vacant home to make it look occupied.[32] Like Picasso's painting, these algorithms integrate a tight feedback loop (hence the term "control" in "control improvisation"). To me, random inputs combined with feedback control and soft, learned constraints look quite promising as a plausible mechanism for creativity.

In practice, all of these "randomized" algorithms are not truly random in that they use pseudorandom sequence generators, which are fundamentally deterministic. But these sequences are not effectively predictable and interact with the inputs to the program in ways that may not be meaningfully different from true randomness. This suggests that even if there is no true randomness in the physical processes of the brain, deterministic chaos could serve the same role.

If the mechanisms used by a computer are randomness plus feedback in a tight embrace, and if the computer accepts, rejects, and tweaks the result based on learned parameters of an AI, then it will be hard to credit the programmer. Tay, the master of supervulgarity, seemed to have developed a personality of her own, not a mere reflection of her creators.

Computers, unlike humans, are algorithmic. Everything they do is carried out as a sequence of discrete steps. The subtle twists of Picasso's brush, in a loop with his visual system and motor neurons, are not algorithmic (see chapter 8). The "decisions" Picasso is making while painting are not discrete events, where he accepts this dab and rejects that drip as individual, atomic actions. The feedback loop that includes random inputs and weighing of alternatives operates in a continuum, in effect yielding an uncountably infinite amount of randomness and appraisal dancing in a tight embrace. In this view, Harris's conclusion becomes glib. It becomes far from obvious that there is no free will.

Computers today are capable of complicated feedback, but for the most part, they are limited by their algorithmic nature and their obsessive determinism. Turing machines cannot perform the dance in a continuum that I just described. There are machines that can, however—our brains. Possibly, other machines will evolve to develop similar mechanisms. In the meantime, computers approximate these capabilities by being extremely fast—bringing an algorithm closer to a continuous process—and by using pseudorandom sequences as a substitute for true randomness.

THE ORIGIN OF SELF

Let us change the question and focus on whether machines have or will ever have "sufficient agency" to hold them accountable for their actions. A key reason that we readily accept the notion that humans have free will is that we have first-person experience that at least seems to us as constituting free will. We intuitively feel that we have sufficient agency. Since other people are built like us, we assume that they have similar first-person experiences. We cannot make any such assumption about machines, but as their complexity increases, will it continue to be safe to assume that they cannot have first-person experiences? To address this question, we have to examine what we mean by a "first-person experience."

Imagine a smooth dome with a perfectly round ball tenuously balanced at the summit (see figure 10.4). If I nudge it, I see the effect; it rolls down the dome toward the north. Or maybe the wind nudges it, and it rolls to the east. But the wind doesn't see the effect; it has no sensory organs to see with. There is no mechanism for the fact of the eastward roll to feed stimulus back to anything related in any way to the production of wind.

But if *I* do the nudging, the situation is completely different. The machine that does the nudging (my brain sending messages to my muscles)

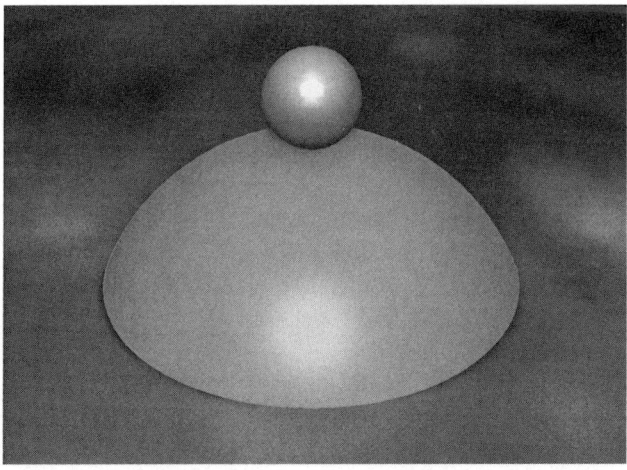

10.4 Ball precariously balanced on the top of a hill.

is connected to sensors (my eyes) that feed signals about the event right back into the machine that caused it. The feedback loop provides the confirmation that creates my sense of self. "I" caused it to roll north, or so it seems to me. What I mean by "I" is that which caused the ball to go north and then sees the ball go north.

If a machine were equipped with a mechanism to produce wind, a camera to "see" the result, and image processing software that can distinguish the ball from the hill and other surroundings, could that machine develop a sense of self? Could it develop something akin to the notion of causation? It would need that notion to develop some form of agency, an "understanding" that its own actions can affect the physical world. Few programs today have the necessary feedback mechanisms (we will see an exception in the next chapter), but I expect many more will acquire them in the future.

HUNEKERS AND MOSQUITOS

A central theme of Douglas Hofstadter's book, *I Am a Strange Loop*, is that feedback loops enclosing a being and its physical environment are essential to the notion of "I." He states in the preface that he considered an alternative title, *"I" Is a Strange Loop*, but rejected that title as "clunky," despite its better representation of the book's topic and goal. Hofstadter uses the term "soul" to mean having an "I," "having a light on inside," "possessing interiority," or "being conscious." This "soul" is the ability to have a first-person experience, to have a sense of self, to have agency. He rejects the idea that this sense of self is an all-or-nothing proposition:

I would like to suggest, at least metaphorically, a numerical scale of "degrees of souledness." We can initially imagine it as running from zero to one hundred, and the units of this scale can be called, just for the fun of it, "hunekers." Thus you and I, dear reader, both possess one hundred hunekers of souledness, or thereabouts.[33]

He then modifies his proposed scale, observing that even humans may have varying "degrees of souledness," and that it changes during our lifetime, starting close to zero when the sperm joins the egg. Hofstadter assigns even a mosquito more than zero hunekers, and suggests the robot vehicles that were around when he wrote his book (in 2007) had at least comparable levels of "perception" to that of a mosquito. A robot vehicle, perhaps, deserves more than zero hunekers. It acts on the world and senses the results, much like a mosquito. If a mosquito were to land on the ball on the hill in figure 10.4 and cause it to roll, would the mosquito somehow "know" that it caused the rolling? Probably, it would "know" no more than a robot vehicle. In both cases, there is feedback, but not complicated enough feedback to create what Hofstadter calls a "strange loop," inducing the sense of "I."

ANNOYING BALLS

Consider staring at the ball perched precariously at the top of the hill in figure 10.4. Suppose that it is just sitting there, seemingly defying gravity, annoying you with its reticence. Will it move? When? You can't bear it anymore, so you reach out and give it a gentle nudge.

How was your conscious mind involved in this decision? Was there "sufficient agency" that I can blame you if the ball rolls over an ant and kills it? Why were you were annoyed at the ball for sitting still?

As you stared at the ball, your brain was generating fake reafferences. Based on your experience with the physical world and your model of that world, you expect the ball to begin rolling at any time, and, as it sits there, it holds you in suspense, violating your expectation. That expectation is a fake reafference, a stimulus that your brain generates and feeds back to itself as the expected stimulus to your senses. You should be seeing the ball roll, but it is not rolling, and the discrepancy is the essence of the feeling of annoyance rising in your brain. That feeling of annoyance is a pattern of neuron firings. These firings are strengthened by the discrepancy between your expectation and what your senses perceive, just like the error signal in the feedback loops of chapter 5. At some point, the neural activity crosses a threshold, triggering a motor efference to nudge the ball. Some time after crossing that threshold, your conscious mind becomes aware of the decision to nudge the ball.

After you nudge the ball, your senses perceive the ball rolling down the hill. This behavior was what your brain expected in the first place, learned from years of observation of the physical world. The error signal attenuates, giving your brain a pleasant peace and contentment, a feeling of satisfaction replacing the prior discord.

Except for the small part in this story about the conscious mind becoming aware of the decision, nothing in this entire process is out of reach for today's computers, augmented with the appropriate sensors. A computer could have "experienced" exactly the same dynamic variation of electrochemical stimulus. If, as Harris says, the conscious mind is outside this loop and largely irrelevant to the story, then you can be held no more accountable than the computer for the crushing of the ant.

What if, before your annoyance reached the critical threshold, a slight breeze arose, and the ball began rolling without your help. You would experience the same release of tension, the same peace and contentment, but without the sense of agency. Events that happen without your intervention do not participate in your sense of self.

If you think about it carefully, however, you didn't *really* cause the ball to roll north, at least not by yourself. Many other causal factors were at least as influential in the outcome. The force of gravity, for example, and the round shape of the dome had at least as much effect on the trajectory of the ball as your nudge. But neither the dome nor the Earth, which produced gravity, were in your feedback loop. You did nothing to cause them. *They* did not cause the ball to roll, your subjective self tells you. You did. How does this subjective sense of agency arise?

When your brain generates the motor efference to cause your arm to reach out, an efference copy is fed back into your very own brain to change the expected sensory stimulus. Now your eyes really *should* see the ball roll. Indeed, if the ball turns out to be glued to the hill, you will experience surprise, a strong neural storm of discord between expectation and perception. If it is not glued to the hill, however, and it begins rolling as expected, the correlation your brain experiences between the efference copy and perception *is* your sense of self. It is this correlation that enables you to tell self from non-self, to sense that it is *your* nudge, *your* finger, *your* action that results in the ball rolling off the hill and crushing the ant.

In chapter 11, I will consider an argument that very much surprised me when I first encountered it. It turns out that quite a few very smart people have argued that the very notion of causation, that an action causes a reaction, is a fabrication of the mind, a consequence of the sense of agency rather than a natural law independent of humans. If these arguments are right, the correlation between efference copies and perception has taught our brains to model the world as actions causing reactions, but in the physical world, there may be no such causation.

A WORM'S SENSE OF SELF

The ability to distinguish self from non-self is present even in much simpler creatures, ones that do not deserve many hunekers. A worm probably doesn't have much in the way of consciousness and probably not much of a model of the physical world, but when it senses that the ground is moving under it, it reacts differently if it is being dragged than if it is contracting and expanding its own body. We will never know what this movement feels like to the worm, or even whether there is anything akin to "feeling like" anything in a worm, but it certainly seems to have the elements of sense of agency. A neural system distinguishes actions initiated by itself from actions initiated by other. There are many devices that already have at least this level of agency, such as the echo cancellers considered in chapter 5. They possess nothing like consciousness, but the elements of most basic form of a sense of self seem to be there.

Even humans can have a sense of agency without the involvement of the conscious mind. You could be riding your bicycle to work, thinking about what's on your calendar, all the while making turns, stopping at red lights, and balancing the bike without involvement of your conscious mind. When you first learned to ride a bike, your conscious mind was very much involved, but you remained a novice as long as your conscious mind stayed in the loop. If you are thinking, "if the bike leans to the left, turn a little to the left to straighten it up," then you are riding very clumsily. The conscious mind is too slow, and we have already seen that feedback loops with substantial delay do not work very well. Your conscious mind has to get out of the way before you can ride the bike well. But when your conscious mind gets out of the way, you do not lose the sense of agency. It is still "you" riding the bike.

When a human learns to play the violin, a similar feedback loop is formed, this time involving the ears rather than the eyes. Practice is essential to tighten this feedback loop to generate pleasant sounds. Such practice is all about training the neural circuitry to bypass the (slow) conscious mind and establish more direct (faster) connections between the auditory circuitry and the muscular control. For both the bicycle and the violin, during the early stages of learning, consciousness is involved. But with practice, the mechanism of feedback control moves from the conscious to the unconscious mechanisms of the brain. It has to be that way because the conscious mechanisms, which are much more complex, have greater delay. Since the time of Harold Black in the 1920s, whom we encountered in chapter 5, control theorists have known that it is difficult to stabilize feedback loops with too much delay.

IS INCOMPETENCE NECESSARY?

Humans (and other creatures) also have feedback loops that we would not associate with any sense of agency. Homeostatic processes such as temperature or pH regulation may be said to be "self-aware" in technical sense. But unlike riding a bicycle, most of us would find it strange to say "I am keeping my body temperature at ninety-eight degrees Fahrenheit." This is not what we mean by "I," but really, what is the difference? My body has temperature sensors and takes actions to correct small deviations from the desired temperature, just as my arms turn the bicycle to the left to straighten it up.

Perhaps, to deserve my one hundred hunekers, I have to pass through the incompetent learning phase, when my conscious mind does a poor job of riding the bicycle, to develop the sense of agency that I subsequently maintain, even after I become an expert bicycle rider. It is still "I" that am riding the bicycle, even if it is not "I" maintaining my body temperature. Both are now automatic, but one was learned with the involvement of my conscious mind, and one was not. The trial-and-error phase, where my "I" probes the physical world, pushes it and watches its reaction, may be essential to the development of a sense of agency.

I will return to this in chapter 12, but it turns out that, for fundamental reasons, *interaction* is more powerful than *reaction*, and reaction alone may not be sufficient to develop a sense of agency. If a worm's ability to crawl without getting confused by the sense of the ground moving under it is simply prewired in its neurons rather than learned, then perhaps it really does not deserve any more hunekers than a thermostat.

We already have software that begins with incompetence and later develops competence by a guided form of trial and error. The backpropagation algorithm considered in chapter 4, which is central to deep learning and used in many AI systems today, starts out hopelessly incompetent. It attempts a classification, then gets its wrist slapped (metaphorically) by a trainer, and it adjusts its parameters to try to do a little better next time. So if incompetence followed by self-improvement is the only requirement for a sense of agency, then AIs are already there.

More likely, with the emergence of digital beings, we may discover that self-awareness can take many forms, some even exceeding anything humans are capable of. For example, we cannot be "aware" in any cognitive sense of the state of every cell in our body. There are simply too many of them. We may be aware of a few at a time, for example experiencing pain when they are being squeezed. But the computers in a server farm are capable of tracking the state of each and every one of the millions of processors. Indeed, server farms are constantly performing such tracking in order to balance their computing load and to identify faulty machines for replacement or repair. In this sense, computers are already capable of forms of introspection that vastly exceeds what humans can accomplish.

SOCIAL CONTRACT

It seems that the question of whether machines can have free will is not very helpful, by itself, for determining whether we can or should hold them accountable for their actions. We are forced to change the question from "do they have free will?" to "do they have sufficient agency that we should hold them accountable for their actions?"

A pragmatic solution may be to limit accountability to humans as a matter of social contract. Our system of justice and law mostly does this, although for some purposes it can also assign accountability to corporations. The US Supreme Court has called free will a "universal and persistent" foundation for our system of law, distinct from "a deterministic view of human conduct that is inconsistent with the underlying precepts of our criminal justice system."[34] The Supreme Court, therefore, dismisses the debate about free will, and simply asserts that accountability is the essential question. They too have changed the question. Otherwise, it would seem that they assume that corporations have free will as well.[35] Their "universal and persistent" principles concern sufficient agency, and whether we call it "free will" is a technicality. The question of whether a corporation has sufficient agency is not an easy question, but it is much easier than whether a corporation has free will.

But limiting accountability to humans and corporations may not be enough. Regardless of whether the world is deterministic, every event in the physical world has been influenced and enabled by countless other events. Where should we pin accountability? Can we continue to limit that accountability to humans and corporations when eventually AIs are designing AIs, and when the original human designers are long dead? What will we do when corporations are run by AIs, or when we incorporate AIs?

Paul Vigna and Michael Casey, in their book on cryptocurrency, describe a scenario, first presented by Mike Hearn at the August 2013 Turing Festival in Edinburgh, where an automated taxi operates with no owner. Vigna and Casey argue that this scenario is plausible once cryptocurrency becomes more widespread.[36]

In Vermont in 1848, an accident with blasting powder blew a thirteen-pound iron rod through the head of Phineas Gage, taking with it much of Gage's prefrontal cortex. Gage, the foreman on a railroad construction crew, survived, but his personality changed. In the words of his friends, Gage was "no longer Gage." Under the materialist stance, we can still hold the pile of biology called Gage responsible for his actions, but is it Gage we are holding responsible? The Stanford neuroendocrinologist Robert Sapolsky, in his 2017 book on the biology of human behavior, documents several other brain disorders, including frontotemporal dementia, Huntington's disease, and strokes, which can change behavior by destroying inhibition.[37] Can patients with these disorders be held accountable for their actions? If so, what about an AI run amuck? And if not, then whom or what do we hold accountable? A glib answer is to pin accountability on the nearest point where a deliberate choice was made, but in many cases, there is no such clear point, and if the world is deterministic, then we can only blame the Big Bang.

According to the Old Testament, God held Eve accountable for eating the apple, putting free will right at the center of Judeo-Christian tradition. In this account, Eve could have done differently, and we would all be still living in Eden. But Eve chose to eat the apple, and here we are, full of sin. Even an all-powerful God was not in control. And yet, if you believe in God, He gets the blame and the credit for all that is.

In the next chapter, I will confront the question whether causation is real. My startling conclusion will be that the very notion of causation is inextricably linked to the notion of a self. Reasoning about causation is impossible without first-person involvement. Since accountability is also inextricably linked to causation, accountability is impossible without an "I" to hold accountable.

11

CAUSES

AUTONOMY

Nobody wants a fully autonomous self-driving car. "Full autonomy" would mean that it would accept no input from humans. It would go where it wants, not where you want. So, how much autonomy should a car have? Obviously, the car, not a human, should decide when to fire the spark plugs (if cars still have spark plugs), but should the car also decide when to apply the brakes? Anti-lock braking systems routinely override driver commands, and most of us appreciate at least that level of autonomy, unless we are stunt drivers. Cooperative cruise control systems adjust the speed of a vehicle to match surrounding traffic. Lane-keeping systems steer the vehicle to keep within a lane on a highway. As car autonomy increases, less is required of the driver until eventually, nothing is required, and the driver becomes a passenger who can sleep, read, or text.

In late November 2018, a couple of California Highway Patrol officers spotted a Tesla Model S electric car going 70 mph down US Highway 101 in Palo Alto at 3:30 a.m. with a man in the driver's seat who was apparently asleep.[1] They chased the car with their lights and siren blaring but got no response. Figuring the car was using Tesla's Autopilot system, for which customers pay an extra $5,000, they called for a backup patrol car to slow the traffic behind them. The backup car created a traffic break, zigzagging across all lanes of traffic to slow and stop the traffic. I once had this done to me, and it's pretty surprising to see a patrol car crossing six lanes of highway back and forth right in front of you. The first patrol car then pulled in front of the Tesla and slowed down. Fortunately, the Tesla did not choose to change lanes and go around them, and they were able to bring the vehicle to a stop. The driver was still asleep. They woke him up by knocking on the window, found him to be drunk, and arrested him for driving under the influence of alcohol.

Does the car or Tesla, the company, bear some of the blame? Other car manufacturers apparently put more effort into preventing this sort of incident. Cadillac's Super Cruise technology uses an infrared camera to measure driver attentiveness and stops the car if the driver appears inattentive. Audi's Traffic Jam Pilot uses an interior gaze-monitoring camera toward the same goal. But perhaps none of this is really necessary, and the car should assume all the responsibility.

In 2014, SAE International, a standards body that was originally established as the Society of Automotive Engineers, published a standard called J3016, *Taxonomy and Definitions for Terms Related to On-Road Motor Vehicle Automated Driving Systems*, which defined six levels of autonomy in vehicles. These levels were based not on the capabilities of the vehicle, but rather on what would be required of a human driver. At level-zero autonomy, the lowest level, the human driver is fully in control, and the vehicle, at most, issues warnings and possibly intervenes momentarily, for example to prevent loss of traction. At level-five autonomy, humans are not involved in steering, accelerating, or braking, although presumably they remain involved in determining the destination; the vehicle is not required to have a steering wheel or brake pedals. A robotic taxi would be a level-five autonomous vehicle. SAE could have defined a level six, where the car would decide where to take you, but they wisely did not.

The notion of autonomy is about causation. If I take a level-five autonomous car to visit my sister, then *I* cause the car to go to where my sister lives, but I do not cause it to apply the brakes at the red lights along the way. If I take a level-zero car to visit my sister, then I cause the car to brake at the red lights, but do not cause the spark plugs to fire.

It is the notion of causation that enables accountability. If a level-zero car hits a pedestrian, the human driver is accountable, not the computers that fired the spark plugs, even though the pedestrian would not have been hit had the computers not fired the spark plugs. If a level-five car hits a pedestrian, accountability is not so clear. Should we hold the car manufacturer or the operator of the robotic taxi company responsible? In either case, do we hold the corporation or the individuals responsible? If it is the individuals, is it the CEO, the engineers, or the marketing people who insisted on level five? What if it is a level-three car, an "eyes off" car that only requires that a human be able to take over in a reasonable amount of time when called upon by the vehicle to do so? What is a reasonable amount of time?

Autonomy, causation, and accountability become ensnarled in ever more difficult questions as technology improves. So-called LAWS, lethal autonomous weapons systems, also called killer robots, are weapons that locate, select, and engage targets without human intervention. Where will we put accountability for their actions? How much autonomy should they be allowed? Is it possible or practical to limit such systems through treaties or laws? If such systems become easy to mass produce, the risks to humanity are incalculable.

None of these questions is easy to answer. The essential question is one of causation. Who or what *causes* the kill? Opposition to LAWS is rooted in trying to ensure that for any kill, we can credibly conclude that it was caused by a human. Today, a human may issue the command to launch a missile, but it will still be the computers that guide the missile toward an infrared heat signature. The determination of ultimate cause gets blurry.

As it happens, the very notion of causation is problematic. It may be a cognitive fiction, in which case, the distinction between robotic causation and human causation may collapse in a heap of nihilism.

A HARMLESS FICTION?

The principle of causality asserts that every effect is produced, as a consequence of some law of nature, by a cause. The notion of making a choice, and hence of free will and accountability, is naturally tied to causality because a choice is meaningless if it does not have some consequence in the world around us. To have a sense of self, we have to have a sense of causation. We (our selves) cause changes in the world around us.

The principle that every effect has a cause has proved surprisingly controversial among philosophers. In 1913, Bertrand Russell challenged the scientific world by denying the very notion of causality:

All philosophers, of every school, imagine that causation is one of the fundamental axioms or postulates of science, yet, oddly enough, in advanced sciences such as gravitational astronomy, the word "cause" never occurs. ... The law of causality, I believe, like much that passes muster among philosophers, is a relic of a bygone age, surviving, like the monarchy, only because it is erroneously supposed to do no harm.[2]

A thoughtful analysis of Russell's position is given by the Caltech philosopher of science Christopher Hitchcock, who concludes that Russell is correct that the notion of causality is often given too much weight and sometimes applied inappropriately, but that Russell overstated the case. Acknowledging the difficulties that Russell was addressing, Hitchcock observes,

It's going to be extraordinarily difficult to find causation within a theory that purports to be universal—about *everything*. Russell's example of gravitational astronomy has just this character.[3]

One obvious reason for the difficulty in such a universal theory is the problem of infinite regress. If *A* causes *B*, what causes *A*? And what causes whatever causes *A*?

The infinite regress problem evaporates, however, with feedback systems like those considered in chapter 5, where we easily have a situation where *A* causes *B* and *B* causes *A* without the emergence of any paradox. Indeed, the University of Pittsburgh philosopher of science John D. Norton observes,

I do not think it is possible to supply a non-circular definition [of causation] and, in practice, that does not seem to matter, since, as I shall indicate in a moment, we are able to apply the notion without one.[4]

Norton uses blob and arrow diagrams to represent cause and effect and points out that such a diagram is incomplete if any blob has no incoming arrow. Hence, the diagram must be infinite in extent or must be circular. Norton's contention is that there is no fundamental difficulty with the diagram being circular.

Objective discussion of causality is difficult because the notion of causality lurks in every aspect of natural language.[5] The very fact that I must use language to convey my ideas to you, and that language has causation built into its very structure, biases my message. Language is causing me to say things I don't want to say, including this sentence.

In physics, causality is often presumed to be even more fundamental than notions of time and space. It has defied efforts to dissect it or find any more primal principles. If we are to talk about what causes an autonomous weapon to kill, we have to try to understand why we do not blame the battery that powers it, and why we might blame the AI it carries. If we could quantify or measure causality, it would help, but this turns out to be surprisingly difficult.

SUBJECTIVE MACHINES

Judea Pearl is an Israeli-American computer scientist who won the 2011 Turing Award for "fundamental contributions to artificial intelligence through the development of a calculus for probabilistic and causal reasoning." His intellectual journey rambled through an impressive variety of topics, starting as an electrical engineer educated at the Technion in Israel. He went to the United States for graduate work, earning a PhD at the Polytechnic Institute of Brooklyn, where he focused on physical devices, particularly superconductors. His career took a turn when he joined UCLA in 1969 and its computer science department in 1970 when the department was formed.

Initially, Pearl's focus was on combinatorial search algorithms, which efficiently explore very large numbers of possibilities. His real passion, however, was studying human cognition, and since 1978 he has called his research group the Cognitive Systems Laboratory. This was never a real laboratory, in the sense of having a space for experiments, unless one counts Pearl's office, which had a permanent sign reading, "Don't Knock. Experiments in Progress." Even when he was working on combinatorial search, his focus was on cognition, or as he said, "the ever-amazing observation of how much people can accomplish with that simplistic, unreliable information source known as *intuition*." As of this writing, well into his eighties, he is still active and contributing brilliant papers and lectures with his charming Israeli accent and disarming humility.

Some readers may know more about Daniel Pearl, Judea Pearl's son, who was a *Wall Street Journal* journalist who was famously publicly executed in

2002 by terrorists in Pakistan. Together with his wife Ruth Pearl, Judea formed a foundation one week after his son's execution that "promotes mutual respect and understanding among diverse cultures through journalism, music, and dialogue," in the words of their website. Judea and Ruth Pearl reacted to personal tragedy in an incredibly generous way, setting out to show the world that Jews and Muslims are not that different from one another and can learn to live together. Together they published a book called, *I Am Jewish: Personal Reflections Inspired by the Last Words of Daniel Pearl*, with a collection of essays on the various ways in which people identify as Jewish.

In many ways, Pearl is not just an electrical engineer and computer scientist, but also a philosopher, a combination that I struggle to emulate. Demurring, in an interview by *3:AM Magazine*, he said,

Philosophers do not consider me one of them. Perhaps because I have degrees in Engineering and Physics or because I show no interest in digging into the irrelevant writings of ancient philosophers.[6]

But his most momentous work is both deeply technical and deeply philosophical. His work with Bayesian networks, also known as belief networks, in his words, had "taught machines to think in shades of gray."[7] Bayesian networks are a technique for reasoning about the relationships between variables that influence one another, but he quickly realized that it was insufficient for reasoning about causes and effects. In his 2018 book, *The Book of Why*, published when he was eighty-two years old, Pearl explains that reasoning about causality cannot be done objectively by looking at data alone. Subjective judgments are essential. This means that to teach machines to reason about causes and effects, we have to teach them to be subjective. We can begin to understand his reasoning by looking at two simple examples.

DOES UGLINESS CAUSE TALENT?

The first example, which Pearl credits to the late Berkeley statistician David Freedman, is the relationship between three variables, the shoe size, the age, and the reading ability of a child. These variables are correlated because children with larger shoes tend to read at a higher level and also tend to be older. If you measure these three variables in some population of children, you will be able to quantify the correlation, but using the data alone, it is impossible to answer a question like "does a larger shoe size cause better reading ability?"

If you are familiar with statistical methods, then you are probably already protesting. Using a standard statistical technique, the correlation between shoe size and reading ability can be eliminated by controlling for age. That is, we partition the data into different age groups, for example to look only at all the children born in January of the same year. We can then separately consider children born in February of the same year, and so on. If our data set is large enough, we will likely discover that in each subset of the data, shoe size and reading ability are uncorrelated, thereby undermining the hypothesis that a larger shoe size causes better reading ability. Haven't we answered the question using the data?

But how did we know to control for age? Pearl's essential argument is that the decision to control for age is based on a background intuition that we have not explicitly acknowledged. It is fundamentally a subjective judgment that the data alone cannot have suggested.

To illustrate this point, Pearl gives another example, which he credits to Felix Elwert and Chris Winship. This example also has three variables, the talent, celebrity, and beauty of Hollywood actors. Admittedly, these features may be hard to measure, so a serious statistical study is unlikely, but nevertheless, the example nicely illustrates Pearl's point. Suppose that in this case we want to answer the question, "does beauty cause talent?"

Suppose that we have collected data about real Hollywood actors and we find that in the overall data set, our measure for beauty is uncorrelated with our measure for talent. This would seem to suggest that the answer to the question is no, beauty does not cause talent. But what if we control for celebrity? That is, we divide our data set so that we consider only the most famous actors, then, separately, the slightly less famous ones, and so on until we have reached the least famous ones. In this case, suppose that the data in each subset shows a negative correlation between beauty and talent. In other words, less beauty is correlated with more talent. This is plausible, Pearl suggests, because celebrity is less likely to arise with either lower beauty or lower talent and extremely unlikely if both are low. But having controlled for celebrity, the answer to our question is a surprising conclusion that ugliness may actually cause talent!

In this case, our background intuition will tell us that controlling for celebrity is the wrong thing to do. But Pearl points out that data alone cannot distinguish these two situations. In one case, we need to control for the third variable, age. In the other case, we need to *not* control for the third variable, celebrity. In fact, using carefully chosen measures for all six variables, it isn't hard to envision a situation where exactly the same numerical data has been gathered for both scenarios. Given exactly the same data, we cannot tell whether we should control for the third variable when assessing the causal relationship between the first two.

Again, if you are familiar with statistical methods, you might be screaming and throwing this book across the room. After all, you have heard the mantra many times, "correlation does not imply causality!" So I shouldn't even be asking the questions of whether beauty causes talent or shoe size causes reading ability. But the reality is that the most valuable uses of statistics are, in fact, asking exactly such questions about causality. Does this drug cure this cancer? If we throw up our hands and rule out asking such questions, we will have dealt a major blow to humanity. Pearl suggests replacing the mantra with "some correlations do imply causation" and then offers a way to figure out which correlations do. His basic argument is that our subjective background intuitions cannot be ignored.

SUBJECTIVE CAUSALITY

Statistics, the practice of collecting and analyzing data, and probability theory, the mathematical methods for reasoning about likelihoods, are often presented as a way of objectively answering questions about the world without letting subjective judgments cloud our reasoning. Let the data speak. But this objective approach, Pearl says, fundamentally cannot reason about causality:

Unlike correlation and many of the other tools of mainstream statistics, causal analysis requires the user to make a subjective commitment. She must draw a causal diagram that reflects her qualitative belief—or, better yet, the consensus belief of researchers in her field of expertise—about the topology of the causal processes at work. She must abandon the centuries-old dogma of objectivity for objectivity's sake. Where causation is concerned, a grain of wise subjectivity tells us more about the real world than any amount of objectivity.[8]

Pearl credits Sewall Wright, who appeared with guinea pigs in chapter 8, for first articulating this principle, and he documents the searing criticisms that Wright had to endure from mainstream statisticians. These critics would accuse Wright of deducing causal relations from correlations. But the critics were misunderstanding Wright. Wright knew full well that we can't deduce causal relations from correlations; every statistician knows this. In a defense of his work, Wright wrote,

> The combination of knowledge of correlations with knowledge of causal relations to obtain certain results, is a different thing from the deduction of causal relations from correlations.[9]

Wright's point is that if we *assume* that there *is* a causal relationship, then a correlation can provide a measure of the strength of that causal relationship.

Pearl's causal diagrams are a powerful tool for reasoning. For the shoe size example, we can draw a causal diagram showing age pointing to both shoe size and reading ability, putting a stake in the ground that we subjectively believe that these are the correct causal relationships (see figure 11.1). Age has a causal effect on shoe size, and it also has a causal effect on reading ability. That is what the diagram says. A statistical study, then, relative to this diagram, can provide a measure of the strength of those causal relationships. The data cannot tell us whether increasing age causes increasing shoe size, but rather can tell us how strong the causality is if we first assume that there is causality. If the data reveal that the causality is weak, then we may conclude that there is no causality. If the

11.1 Two causal diagrams reflecting subjective judgments about causal relationships. After Pearl and Mackenzie, *The Book of Why* (2018).

data support the hypothesis that the causality is strong, however, then the same data would support the opposite hypothesis, that increasing shoe size causes increasing age. It is only our background intuition that blocks that conclusion, not the data.

For the second example, the arrows are reversed, as shown in figure 11.1. Talent points to celebrity and beauty points to celebrity, suggesting there is a causal relationship between talent and celebrity and between beauty and celebrity. Again, such a diagram represents a subjective judgment, and given the diagram as a framework, a statistical study can be used to measure the strength of the presumed causal relationships. We can measure how strongly beauty causes celebrity, for example.

CONFOUNDERS AND COLLIDERS

We can use causal diagrams as background information to study two hypotheses, that shoe size causally affects reading ability, and that beauty causally affects talent. These two hypothesized causal relationships are shown in figure 11.2 as dashed arrows with a question mark. But how should we use the data to measure these causal relationships? What variables, if any, should we control for? If we fail to control for age, then the data will support the hypothesis that shoe size causes reading ability. It is a statistical mistake to fail to control for age. In the other example, however, it is a statistical mistake to control for celebrity. If we control for celebrity, then the data will support the hypothesis that ugliness causes talent.

Pearl's technique is astonishingly easy to use, even if understanding why it works requires some careful thinking. Pearl calls the pattern shown on the left in the two figures, where one variable (age) points to

11.2 Two hypotheses about causal relationships that can be tested using the background information of figure 11.1.

two others (reading ability and shoe size), a "confounder." Specifically, age is a confounder for the relationship between reading ability and shoe size because it has an (assumed) causal effect on both. Pearl calls the pattern on the right, where two variables (talent and beauty) each point to a third (celebrity), a "collider." Celebrity is a collider on the relationship between talent and beauty.

The rule is then very simple. Control for confounders. Do *not* control for colliders. The beauty of Pearl's technique is that it forces us to make our background assumptions explicit in a diagram and then gives a simple formulaic way to use the diagram to figure out how to correctly handle the data. The conclusions we draw, of course, will only be valid to the extent that our assumptions about causal relationships are valid. And data alone cannot tell us whether these assumptions are valid. For the questions represented by the dashed arrows in figure 11.2, if we assume the solid arrows are valid and use those solid arrows to figure out whether to control for the third variable, then the data will tell us that the dashed causal relationship is weak or nonexistent in both cases.

What does it mean for the assumed solid arrows to be valid? If Bertrand Russell is right, it actually means nothing, at least in an objective, physical sense. But in the context of an individual harboring the first-person illusion (if it is an illusion) of being able to take action, it means a great deal. To our first-person selves, actions have consequences. But the causal relationships, and hence any accountability we assign, are fundamentally subjective.

Causal analysis is emphatically not just about data; in causal analysis we must incorporate some understanding of the process that produces the data, and then we get something that was not in the data to begin with.[10]

If we do not have some prior understanding of the process that produces the data, then observing the data alone will tell us nothing about causality. This does not bode well for machines if we expect them to reason about causality. It suggests that they will have to achieve some level of understanding and subjectivity before they will be able to handle the notion of causality. If machines are unable to reason about causality, then how can we hold them accountable for causing anything? I will argue next that they can, in fact, reason about causality, and indeed, they already do.

HYPOTHETICAL INTERVENTION: COUNTERFACTUALS

Some measure of understanding and subjectivity may not be so out of reach for machines. They can learn about causality by *intervening* rather than just observing. If they can *interact* with the system being studied, then it becomes possible to reason about causality. Interaction is more powerful than observation.

Intervention has its limitations in practice, however. For the celebrity example, we can't just set the talent of actors to whatever level we want and then measure their celebrity. For such a case, we can reason using counterfactuals. A counterfactual is simply a statement of something that is not the case. We can say, for example, "if Sophia Loren were less beautiful, she would be less of a celebrity." This is a counterfactual because Sophia Loren is not less beautiful. An experiment that involved intervention, making her less beautiful, would probably be unethical, so we should avoid going there. Hence, the causal relationship from beauty to celebrity will probably forever remain unverified, at least by this technique. But the counterfactual helps us build confidence in the intuition, and once we have confidence in that causal relationship, we can use this assumption to study other causal relationships, such as between beauty and talent.

Judea Pearl's own work has been motivated by the goal of getting AIs to exhibit more humanlike intelligence:

Artificial intelligence (AI) researchers...aimed to build robots that could communicate with humans about alternate scenarios, credit and blame, responsibility and regret. These are all counterfactual notions that AI researchers had to mechanize before they had the slightest chance of achieving what they call "strong AI"—humanlike intelligence.[11]

Counterfactuals can be understood as a form of hypothetical intervention, like the fake reafferences that we considered in chapter 10. According to Pearl, even though they involve variables that can never be observed in real life, they nevertheless give us a powerful reasoning tool. But they still require subjective interpretation.

ACTUAL INTERVENTION

In many cases, we can do better than hypothetical intervention. Suppose that we are interested in evaluating whether a particular drug is effective against some disease. In other words, we wish to measure the strength of a hypothesized causal relationship from treatment (whether a treatment is administered) to some measure of health (see figure 11.3). Suppose that there is risk of a confounder, some factor that causes a patient to be more or less likely to take the treatment and also affects the patient's health. The confounder could be, for example, gender, age, or genetics. To be specific, suppose that the treatment for some disease is more appealing to women than men, and that women tend to recover more from the disease than men. In that case, gender is a confounder and failing to control for it will invalidate the results.

In many cases, however, we don't know what confounders might be lurking in the shadows, and there may be confounders that we cannot measure. There might be some unknown genetic effect, for example. We can't control for confounders that we can't measure or that we don't know exist. Is it hopeless, then, to evaluate whether a treatment is effective?

To guard against the risk of hidden confounders, Pearl points out, active intervention is effective (when active intervention is possible), underscoring that interaction is more powerful than observation alone. We must somehow force the treatment on some patients and force the lack of treatment on others. This could be done by an act of will on the part of a doctor, resulting in the diagram shown on the right in figure 11.3. This is what Pearl calls a "do-operator." Do something (administer a drug) and watch the reaction. If the act of will is genuinely free, and specifically cannot in any way be influenced by the confounding factor, then this act of will breaks the causal relationship between the confounding factor and whether the treatment is administered. Controlling for the confounding factor is no longer necessary.

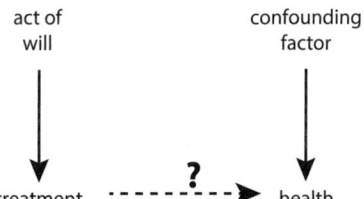

11.3 A causal diagram on the left guiding the evaluation of a treatment's effectiveness that requires controlling for a confounding factor. On the right, an act of free will removes the effect of the confounder.

There is still a potential pitfall, however. How can we ensure that the act of will is genuinely free from influence by the confounding factor? Perhaps the doctor is more likely to choose women to receive the treatment than men, in which case, the confounder has not really been removed. The standard way to guard against this is to use random choice rather than an act of will. It is fascinating to me that an act of truly free will can be replaced by a random act when it comes to reasoning about causality. This suggests that the relationship between nondeterminism and free will is more subtle than Sam Harris implies (see chapter 10).

In the next chapter, I will show that an act of free will is not quite equivalent to a random choice. If a doctor uses free will to choose who gets the treatment, then the results of the trial will always be suspect. Observers can always have lingering doubt that the biases of the doctor affected the choices. Only the doctor, as a first-person self, will be sure of the validity of the trial. If on the other hand the doctor chooses who receives the drug by publicly flipping a coin, then any lingering doubts about bias vanish.

RANDOMIZED CONTROLLED TRIALS

Randomized controlled trials (RCTs) are the gold standard for determining causality in medical treatments and many other problems. The way an RCT works is that a pool of patients is selected, and within that pool, a randomly chosen subset is given the drug and the rest are given an identical looking placebo. Ideally, both the patients and the medical personnel are unaware of who is getting the real drug, and the choice is truly random, unaffected in any way by any characteristic of the patients. Like an act of will, this is a form of intervention because we have forced the value of one of the variables, whether the drug is taken, for each of the patients.

The reason that an RCT works is that, in a causal diagram, it eliminates all the incoming arrows to the variable that says whether or not a patient took the drug (see figure 11.4). This variable is unaffected by other variables such as the gender, age, or genetics of the patient. Hence, it eliminates any confounders between taking the drug and the health of the patient. Of course, the constitution of the pool matters; if, for example, the pool consists only of men, then the study will not tell us anything about whether the drug works for women.

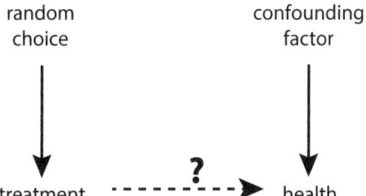

11.4 A causal diagram for the evaluation of a treatment's effectiveness using an RCT.

RCTs are actually routinely used in software today. It is common at Facebook, for example, when considering a change to the user interface, to randomly select users to whom a variant of the user interface is presented. The reactions of the users, whether they click on an ad, for example, can be measured and compared to a control group, which sees the old user interface. In this way, Facebook software can determine whether some feature of a user interface causes more clicks on ads. This process can be automated, enabling the software to experiment and learn what causes users to click on ads. This is a much more powerful form of reasoning than mere correlation, and it can result in software designing and refining its own user interfaces. The software can even learn to customize the interface for individual users or groups of users.

Humans do this too, albeit in a less disciplined way. We experiment with our "user interfaces," how we interact with other humans, and (hopefully) improve it over time. We might try fist bumping instead of shaking hands, for example, or making our handshake firmer or weaker. If we pick the people with whom we try something new at random, then we are performing an RCT. Most of us will not be very good at picking our subjects at random, however. We are unlikely to fist bump the CEO of the company we work for, for example.

It is not always possible or ethical to conduct an RCT. Pearl documents the decades-long agonizing debate over the question of whether smoking causes cancer. Had it been possible or ethical to randomly select people and make them smoke or not smoke, the debate may have been over much earlier. Instead, we were stuck with correlations and hypotheticals.

UNCAUSED ACTION

If causal reasoning is necessarily subjective, then was Russell right that physics has no need for causality? Classical physics, as defined by Sir Isaac Newton in the seventeenth century, is permeated with the concept of force and deterministic reactions to force. Despite Russell's challenge, most people do interpret classical physics in a causal way. A force causes an object's motion to change. In fact, the very meaning of "force" is the cause for the change. This classical theory is also supposed to provide a deterministic framework for the universe where everything has a cause and each cause has a unique effect. But this theory has holes in it that admit both nondeterminism and uncaused effects.

A nice example is due to the philosopher of science John Norton.[12] Norton considers the same physical system, a ball precariously balanced on top of a hill, that appeared in the previous chapter (see figure 10.4). Norton points out that, without violating any of Newton's laws, the ball can spontaneously begin rolling off the hill in an arbitrary direction at an arbitrary time without anything causing it to start rolling. His argument is carefully constructed and surprising to anyone who has studied physics (or at least it was surprising to me when I first heard it).

Newton's second law states that at any time instant, the force imposed on an object equals its mass times its acceleration. If there is no force, the acceleration must be zero. If the acceleration is zero, then the velocity is not changing. Hence, it would seem, that if the ball is not moving, balanced on the top of the hill, and no force is applied, then it should remain still, with velocity equal to zero. But Norton pointed out that it is possible for the ball to start rolling down the hill at any arbitrary time T without violating this law and without any force initiating the roll. At the instant T, the ball will have velocity zero and acceleration zero, so it is not moving. But at any time greater than T, say at $T+\varepsilon$, no matter how small ε is, the ball may be no longer centered on the top of the hill. It will now be sitting on a slope, which means that gravity will exert a nonzero force in the downhill direction, and the ball will have a nonzero acceleration.

Let me repeat because, in my experience, even experts in physics can have trouble grasping what is happening at time T. At the instant T, the ball is not moving, the net force on the ball is zero, and the ball is not accelerating. At any time larger than T, the ball is moving, the net force is not zero, and the ball is accelerating down the hill. This can occur at any time T without violating Newton's second law. The law holds at every instant.

To help clear any confusion, Norton says,

We are tempted to think of the instant $t = T$ as the first instant at which the mass moves. But that is not so. It is the *last* instant at which the mass does *not* move. There is no first instant at which the mass moves.[13]

Norton should be saying that there is no first instant at which the mass *accelerates*, so there is no instant at which an external force is required to start the motion. But we get his point. At all instants greater than T, there is a nonzero external force, gravity on a slope, so the mass accelerates.

What about Newton's other laws? Are they violated when the ball spontaneously starts rolling? Newton's first law states that an object will remain at rest or in uniform motion in a straight line unless acted upon by an external force. It may seem that having the ball spontaneously start to move violates this law, but actually it does not. Since an external force may vary in time, this first law needs also to be interpreted as a statement that holds at each instant of time. And indeed, with Norton's example, it holds at every instant of time. Assuming T is the latest time at which the ball is still, then at all times $t \leq T$, there is no net external force on the ball, and the ball sits still. At all times $t > T$, there is a nonzero net external force on the ball, so the ball accelerates.

What about Newton's third law, which states that every action has an equal and opposite reaction? Here, we have to understand something peculiar about gravity. When gravity exerts a force on a ball, causing the ball to fall, the ball exerts an equal and opposite force on the Earth, causing the Earth to rise. The mass of the Earth, however, is so much larger than that of the ball that, to the Earth, the force exerted on it is negligible. The resulting acceleration of the Earth will be too small to measure. When a mosquito lands on you, it pushes you, but you do not fall over.

Once again, we have to interpret the third law as applying at all instants in time. At all times $t \leq T$, gravity pulls the ball straight down, but the rigidity of the hill pushes it back up, so the net force on the ball is zero. Correspondingly, the net force on the Earth is zero, so the Earth does not rise. But at all times $t > T$, the nonzero net force downward on the ball will be balanced by a tiny nonzero net force pulling the Earth up. The effect of that force will not be measurable, but it is there, so the third law is also not violated.[14]

METASTABLE STATES

Neither computers nor biological creatures are made out of balls on hills. Their digital circuits and neurons, however, are vulnerable to similarly uncaused action. When the ball is perched on the top of the hill, it is in what is called a metastable state. Such a state is technically stable, but on closer examination the ball is vulnerable to falling out of that stable state with no provocation. Digital circuits, particularly ones at the boundary between the continuous physical world and the discrete world of digital electronics, have long been known to be vulnerable to lingering for unbounded periods of time in a metastable state.[15] So the problem is not limited to cute examples of balls on hills. It is a fundamental problem at the boundary between the discrete, computational world of computers and the continuous physical world.

Metastable behavior has also been observed by biologists in nerve axons, which, under certain circumstances, can have two resting potentials.[16] It is plausible, therefore, that metastability plays a role in brain function.

DETERMINISM

The notion of causation is hard to separate from the notion of determinism. Broadly, determinism in the physical world is the principle that everything that happens is inevitable, preordained by some earlier state of the universe, and then following from the laws of physics. John Earman, in his *Primer on Determinism*, echoes the difficulty that many thinkers have had with the concept:

This is already enough to make strong the suspicion that a real understanding of determinism cannot be achieved without simultaneously constructing a comprehensive philosophy of science. Since I have no such comprehensive view to offer, I approach the task I have set myself with humility. And also with the cowardly resolve to issue disclaimers whenever the going gets too rough. But even in a cowardly approach, determinism wins our unceasing admiration in forcing to the surface many of the more important and intriguing issues in the length and breadth of the philosophy of science.[17]

But Earman insists that "determinism is a doctrine about the nature of the world." Determinism is much easier to understand if we define it instead as a property of *models* of the physical world, not a property of the physical world itself.[18] We can define determinism of models as follows:

A model is deterministic if given an initial *state* of the model, and given all the *inputs* that are provided to the model, the model defines exactly one possible *behavior*.

In other words, a model is deterministic if it is not possible for it to react in two or more ways to the same conditions. Only one reaction is possible. In this definition, the italicized words have to be defined within the modeling paradigm to complete the definition, specifically, "state," "input," and "behavior." It also requires that the model have a notion of an "input," which already requires a notion of causation.[19]

For example, if the *state* of a ball is its position in a Euclidean space at a time *t*, the *input* is a force applied to the ball at each instant *t*, and the *behavior* is the motion of the ball through space, then Newton's second law may seem to provide a deterministic model that tells us how the ball will move. But as we have seen, the model isn't really completely deterministic.

FORCE

The concept of force and its association with causality are central to classical physics. In his wonderful chronicle of the philosophy of the Vienna School, Karl Sigmund quotes the physicist Ernst Mach, one of founders of the field of philosophy of science:

Let us direct our attention to the concept of force.... Force is a circumstance leading to movement.... The circumstances giving rise to movement that are best known to us are our own acts of volition, the results of nerve impulses. In the movements that we ourselves initiate, we always feel a push or a pull. From this simple fact arose our habit of imagining all circumstances that give rise to movement as akin to volitional acts, and thus as pushes or pulls...

Whenever we attempt to discount this conception [of force] as subjective, animistic, and unscientific, we invariably fail. Surely it cannot profit us to do violence to our own natural thoughts and to deliberately inhibit our minds in this regard.[20]

Sigmund sums up Mach's radical view as, "causality is nothing but the regular connection of events. In this sense, causal links do not provide an additional 'explanation.'" The notions of causality, volition, force, and free will are thereby inextricably entangled, and all may be mental constructs, not objective facts about the world. Does this put these notions out of reach for machines?

CHOICE

When I slow down my car or push open a door, do I *choose* to do these things? What does it mean to make such a choice? Hitchcock has also observed that the notion of causation is tied up with free will and determinism. Citing the philosophers David Hume and R. E. Hobart (a pseudonym for Dickinson Miller), he says,

Freedom would be *undermined* if there were no connection between our choices and their physical outcomes. Problems arise, however, when we try to bring the decision-making process itself within the scope of the deterministic theory.[21]

As I explained in the previous chapter, Daniel Dennett resolves these problems rather nicely by asserting that free will itself is a human cognitive construct, much the way Mach asserts that the concept of force and the notion of causality are human cognitive constructs. Further confusing the issue, although causation is required for free will to make any sense (our actions must cause consequences), freedom to choose means, to many people, that the choice itself is not caused by prior conditions. Patricia Churchland, a philosopher at UC San Diego who reexamines philosophical questions in light of what has been learned in neuroscience, in her book, *Touching a Nerve*, explains this point of view nicely:

You can mean that if you have free will, then your decisions are not caused by anything at all—not by your goals, emotions, motives, knowledge, or whatever. Somehow, according to this idea, your will (whatever that is) creates a decision by reason (whatever that is). This is known as the contracausal account of free will. The name contracausal reflects a philosophical theory that really free choices are not caused by anything, or at least by nothing physical such as activity in the brain. Decisions, according to this idea, are created free of causal antecedents. The German philosopher Immanuel Kant (1724–1804) held a view roughly like this, and some contemporary followers of Kant do also.[22]

Churchland does not put much currency in this contracausal account of free will, saying "this is an idea espoused mainly by academic philosophers, not by dentists and carpenters and farmers." She offers a more common-sense definition:

If you are *intending* your action, *knowing* what you are doing, and are of sound mind, and if the decision is not coerced (no gun is pointed at your head), then you are exhibiting free will. This is about as good as it gets.[23]

Daniel Dennett also dismisses the contracausal interpretation:

Free will isn't what some of the folk ideology of the manifest image proclaims it to be, a sort of magical isolation from causation. ... I wholeheartedly agree with the scientific chorus that that sort of free will is an illusion, but that doesn't mean that free will is an illusion in any morally important sense. It is as real as colors, as real as dollars.[24]

However, if causation itself is (only) as real as colors and dollars, as suggested by Russell, then Churchland's and Dennett's common-sense free will is completely consistent with causation. Neither is required to exist in the physical world except as mental states. And Dennett implies that the existence of free will matters because of its role in a moral system, itself another collection of mental states. Perhaps all we can say is that free will, causation, and morality together form a self-consistent and effective collection of dynamic mental states, illusions that enable us to live our lives. But this sanguine peace is about to be disrupted by the intrusion of AIs with rich enough feedback mechanisms to endow them too with notions of causation and enable them to take actions for which we may have to hold them morally accountable.

DETERMINISM AND INTERACTION

John Norton's example of a ball on top of a hill that can start rolling at any time illustrates that not only do Newton's laws not require that an action have a cause, but they also do not give us a deterministic model. Given the same initial state (the ball is on the top of the hill) and the same input (no force is applied except gravity), many behaviors are allowed by the model.

Nevertheless, our best scientific understanding of the physical world is mostly, albeit not completely, deterministic. In the early 1800s, during the heyday of Newtonian mechanics, the French scientist Pierre-Simon Laplace argued that if some "great intellect" (later named "Laplace's demon") were to know the precise location and velocity of every particle in the universe, then the past and future locations and velocities for each particle would be completely determined and could be calculated from the laws of classical mechanics.[25] As Norton's example shows us, even without more modern physical theories such as quantum mechanics and relativity, Laplace overstates the determinism of Newtonian mechanics. Newtonian mechanics does, in fact, admit nondeterminism.[26]

What about the probabilistic nature of the wave function in quantum mechanics? Does this undermine the idea of a deterministic universe? The late physicist Stephen Hawking argues that it does not:

> At first, it seemed that these hopes for a complete determinism would be dashed by the discovery early in the twentieth century that events like the decay of radioactive atoms seemed to take place at random. It was as if God was playing dice, in Einstein's phrase. But science snatched victory from the jaws of defeat by moving the goal posts and redefining what is meant by a complete knowledge of the universe.[27]

Hawking is referring to the fact that the Schrödinger equation, which describes how a wave function evolves in time, is deterministic. In effect, using quantum theory, we can redefine *state* and *behavior* in our model of the universe and turn a nondeterministic model into a deterministic one. Hawking sums up, "In quantum theory, it turns out one doesn't need to know both the positions and the velocities [of the particles]." It is enough to know how the wave function evolves in time. Position and velocity no longer represent the state or the behavior of a system.

However, the Schrödinger equation describes how an *isolated* system evolves in time. Quantum mechanics has a nasty "observer problem," where the mere presence of an outside observer changes the behavior of the system.[28] Quantum physics tells us that observation is impossible without interaction. You cannot get outside the system and watch it. Not only is interaction more powerful than observation, but observation alone cannot be done! The objectivity of dispassionate statistical inference unblemished by properties of the observer is not physically achievable. Subjectivity rules!

DETERMINISTIC MACHINES IN A RANDOM WORLD

In the previous chapter, I pointed out that determinism is about *when* the resolution between alternatives is made, not about why or how. In the extreme case of a hypothesized deterministic universe, alternatives are resolved at the time of creation, or at the Big Bang. This extreme form of determinism, however, is inconsistent with our best understanding of physics today.

Nevertheless, humans have gone to great lengths to make computers deterministic. We would not entrust them with the world's banking were it not for their almost perfectly deterministic operation. But "almost perfect" is not the same as perfect. Moreover, as machines become more embodied (see chapter 7), they will incorporate in their operations a greater diversity of physical processes, letting more nondeterminism into the tent.

The Austrian-British philosopher of science Karl Popper (1902–1994), a champion of objectiveness and of a deterministic universe, argues that this inability to predict an outcome is not a consequence of intrinsic randomness in the world but rather a consequence of our lack of knowledge of all of the details of the underlying physical system.[29] But this begs the question of whether such knowledge is *possible*. If we strive to obtain this knowledge through measurement of physical conditions, then we will be subject to the limitations of Shannon's channel capacity theorem (see chapter 8).

Noting that interaction is more powerful than observation, as AIs acquire more ability to intervene in their environment, their learning algorithms will acquire the ability to reason about causality, as done for example in the Facebook user interface. Pearl requires this if he is to consider the machine intelligent:

> To me, a strong AI should be a machine that can reflect on its actions and learn from past mistakes. It should be able to understand the statement "I should have acted differently," whether it is told as much by a human or arrives at that conclusion itself.[30]

A thermostat should be able to learn that it causes the temperature fluctuations in a room. This would endow the thermostat with better self-testing and enable it to shut itself down if the room is miswired (see chapter 9). Indeed, there is a fledgling field concerned with "self-aware systems," with a number of workshops worldwide addressing the question of how to design software that gathers and maintains information about its current state and environment, reasons about its behavior, and adapts itself as necessary. Active intervention in the environment is an essential tool for such systems. And active intervention is a first-person activity, not a third-person observation. In the next chapter, I examine in more depth the relationship between interaction and the notion of a first-person, subjective self.

12

INTERACTION

INTERACTION RULES!

In this chapter, I would like to share with you some insights that I got while pondering the following question: Does it matter whether machines have a sense of self, the ability to engage in an interaction as a first-person participant? Does it matter whether they have an "I" in the same sense that you and I do (assuming you are a human reader)? If they are able to do everything I can do without such a sense of self, then, arguably, it does not matter. But in pondering this question, I stumbled on some technical results that demonstrate that there are things that cannot be accomplished without first-person involvement in an interaction. These results can, in fact, be used to give a technical meaning to "first-person interaction," a concept that philosophers have struggled with for centuries. These results may help us pin down the elusive "I."

The insights that I would like to share depend on some rather technical concepts from computer science. I will do my best to explain these concepts, but if your tolerance for the technical is low, feel free to skip the rest of this chapter. The next chapter comes back down to earth. The bottom line here is that first-person interaction can accomplish things that remain inaccessible to a third-person observer. For example, I will show that without first-person interaction, it is not possible to distinguish an entity that cannot have free will from one that can. I will also show that free will and random choice are *almost* interchangeable, but not quite. The difference between the two is a first-person difference, where an "I" with free will can gain information that no third-party observer can gain, and that merely replacing that free will with random choice makes the information available to a third-party observer. Bear with me. The concepts are subtle and deep.

Interaction combines observation and action in a closed feedback loop. We have already seen evidence of the power of interaction in a surprising variety of ways. In chapter 4, we saw that the biggest breakthroughs in AI, which came to maturity in the last ten years, replaced the prior open-loop good old-fashioned AI (GOFAI) techniques with feedback. Deep learning is fundamentally a feedback technique, and it yields results of such complexity as to be inexplicable, as we saw in chapter 6.

In chapter 5, we saw how Harold Black, way back in the 1920s, found that feedback could compensate for deficiencies in feed-forward circuits. His feedback circuits push on their environment, measure the extent to which its reaction deviates from the desired reaction, and adjust the pushing to get closer. Today, Black's feedback principle is used in many engineered systems. It makes compact loudspeakers sound good and enables smart speakers to hear you over the music they are playing. It keeps planes in the air, shoots planes out of the air, and keeps car engines running smoothly. It prevents brakes from locking, cars from crashing, and pipes from freezing. It keeps stocks from crashing and causes stocks to crash. In biological systems, feedback shapes bones, makes intelligible speech, and makes it possible to distinguish self from non-self.

In chapter 7, we explored the thesis of *embodied cognition*, where the mind "simply does not exist as something decoupled from the body and the environment in which it resides."[1] The mind does not just interact with its environment, but rather the mind *is* an interaction of the brain with its environment. A cognitive being is not an *observer* of its environment, but rather a collection of feedback loops that include the body and its environment, an *interactive* system.

In chapter 10, interaction came together with randomness and free will in the discussion of Picasso's painting, *Les Demoiselles d'Avignon*. Here, embodied cognition seems obvious, where the paintbrush, the canvas, and the drips of the paint, with their intrinsic randomness, coalesce with Picasso's brain, delighting it with accidents in a tight feedback embrace that produced one of the world's most famous works of art.

In chapter 11, we saw the thesis that the very notion of causality may stem from interaction of a mind with its environment, not from fundamental physics. We heard Judea Pearl's argument that interacting with a system enables drawing conclusions about causal relationships between pieces of that system, conclusions that are much harder to defend without interaction. Here too, interaction came together with randomness (in a randomized controlled trial [RCT]) and acts of will (Pearl's "do-operator"). In this chapter, I will show you that this relationship between randomness and acts of will is fascinating and subtle. They are not completely interchangeable. This follows from a beautiful idea known as a zero-knowledge proof.

ZERO-KNOWLEDGE PROOFS

Consider an interaction between two parties, Shah Fi and Mick Ali. Shah knows something important, like a password, and wants to prove to Mick that she knows this. She is a very private person, so while Shah wants to convince Mick that she knows the password, she does not want Mick to be able to convincingly tell anyone else that she knows the password. Her objective is only to convince Mick and give him exactly zero additional information. Note that Shah Fi's objective cannot be accomplished by simply telling Mick the password because then Mick will then also know the password.

One way to accomplish this is using the notion of a zero-knowledge proof. This notion is easy to understand using a story developed by Jean-Jacques Quisquater and Louis Guillou, in collaboration with their children, who are listed by their first names as coauthors on the paper.[2] In this story, there is an oddly shaped cave (see figure 12.1), where the entrance tunnel forks into two tunnels labeled *A* and *B*. Both tunnels are dead ends, but there is a door connecting the two ends. The door can only be opened with a password that Shah knows.

One way that Shah could prove to Mick that she knows the password is to enter the cave together with Mick, and while Mick waits at the mouth

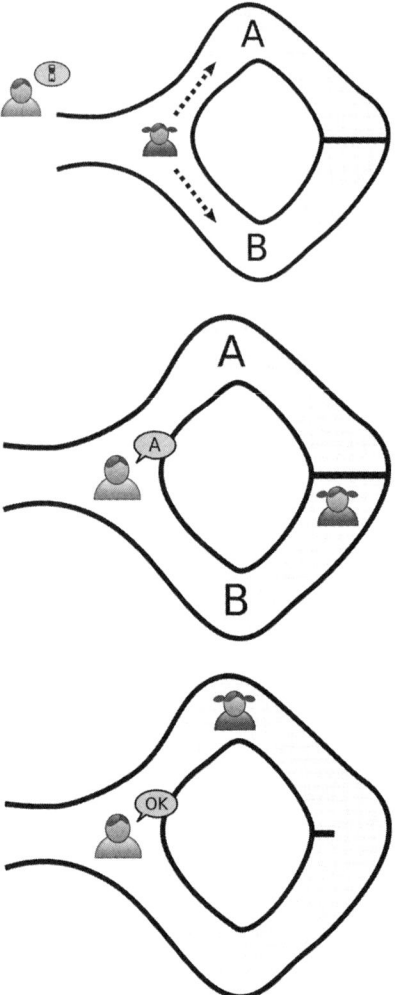

12.1   Ali Baba's cave, illustrating zero-knowledge proofs. After drawings by Dake, via Wikimedia Commons CC BY 2.5.

of the cave, go down tunnel *A* and come back out through tunnel *B*. Mick will be convinced that Shah knows the password, and Mick will not himself know the password. But if Mick surreptitiously records the event with a video camera, then Mick would be able to convince anyone else that Shah knows the password. This gives Mick more power than Shah wants to give him. She really does not trust this guy!

By going in one way and coming out the other, Shah Fi is providing convincing proof that she knows the password. But it is a proof that is available to any passive observer. This is why it is sufficient for Mick to make a video in order to convince others. The proof does not require any interaction with Shah Fi. But if we add interaction, then Shah Fi can make it much more difficult for Mick to share his knowledge.

So, instead, Mick waits outside the cave while Shah goes in and picks one of the tunnels to go down. Suppose she picks tunnel *B* and goes as far as the door. Then Mick comes into the cave as far as the fork and randomly calls out either *A* or *B*. He cannot see which tunnel Shah went down. If he calls *A*, then Shah has to use her password, open the door, and come out through tunnel *A*. Mick is not yet *sure* that Shah knows the password, but he can conclude that it is equally likely that she knows it as that she doesn't know it.

Mick and Shah then repeat the experiment. If Shah successfully comes out of the tunnel that Mick identifies a second time, then Mick can conclude that the probability that she knows the password is now 3/4. It would have required quite a bit of luck for her to not have to use the password twice in a row. Repeating the experiment again will raise the probability to 7/8, or, equivalently, the probability that she got lucky and didn't need the password has dropped to one in eight. After ten repeats, the likelihood that she didn't need the password drops to about one in one thousand. By repeating the experiment, Shah can convince Mick to any level he demands short of absolute certainty.

Unlike the previous experiment, where Shah just went in one tunnel and came out the other, this new experiment does not give Mick the power to convince a third party, say Char Lee, that Shah knows the password. Mick could videotape the whole experiment, but Char is a savvy third party, and he suspects that Mick and Shah colluded and agreed ahead of time on the sequence of *A*'s and *B*'s that Mick would call out.

Only Mick and Shah can know whether collusion occurred. So Char is not convinced that Shah knows the password the way Mick is convinced. Shah retains plausible deniability, and only Mick knows for sure (almost for sure) that she knows the password.

LET'S GET PERSONAL

There are several fascinating aspects to this story. First, for Shah to prove to Mick that she knows the password while not giving Mick the power to pass on that knowledge, interaction is required. If Mick simply watches Shah, observing but not interacting, then anything Shah does to convince Mick that she knows the password gives Mick the power to pass on that knowledge, for example by making a video. He can then convince Char Lee that Shah knows the password by simply showing him the video. But by interacting, Shah is able to convince Mick and only Mick. No third-party observer will be convinced. You have to actively participate to be convinced. Interaction is more powerful than observation, but for interaction to work, you have to be a first-person participant in the interaction. This is what interaction means! Mick's first-person action, choosing A or B at random, is necessarily subjective. Only he knows that no collusion was involved. As with Pearl's causal reasoning of the previous chapter, subjective methods can accomplish something that no objective method can.

Another fascinating aspect of this story is the role of uncertainty. Using this scheme, it is not possible to give Mick absolute certainty without giving Mick more than Shah wants to. The residual uncertainty that Mick retains can be made as small as we like, but it cannot be reduced to zero, at least not by this technique.

A third fascinating aspect is the role of randomness. Mick has to know that the sequence of A's and B's that he calls out are not knowable to Shah (with high probability), but that fact has to be hidden from anyone else. Mick could choose A or B each time using his free will, if he has free will. Actually, all that is required is that Mick *believe* that he has free will and believes that he has chosen randomly between A and B. Given this belief, he will be convinced that with high probability Shah knows the password. It makes no difference whether the choice is made by Mick's conscious mind or by some unconscious mechanism in his brain. If he believes he has free will, every aspect of the system plays out exactly as if he has free will. This supports Daniel Dennett's perspective on free will and refutes that of Sam Harris (see chapter 10).

Suppose that Mick chooses instead to rely on an external source of randomness rather than some internal free will. He could, for example, flip a coin each time to choose between A and B. But this could result in leaking information because now he could videotape the coin flipping, and the resulting video would convince Char Lee and any other third party as much as it convinces Mick. It will be evident to any observer that Mick is not colluding with Shah. To preserve Shah's privacy, Mick has to generate the choices between A and B in a hidden way, and by hiding this, he gives up the ability to convince any third party.

As discussed in the previous chapter, Judea Pearl observed that an act of will, the do-operator, is interchangeable (more or less) with random choice, a randomized controlled trial. The Mick-Shah story shows us that for an RCT to be effective, every observer must believe in the randomization, must believe that there is no collusion. Unlike Shah's situation, the whole point of an RCT is to convince *everybody*, whereas Shah wants to convince Mick and only Mick. The easiest way to convince everybody is to make visible the mechanism used to obtain the random choices, for example by coin flips. An act of free will will not be visible and therefore should not be trusted as a replacement for an RCT. Otherwise, the only person that will be convinced by the trial is the person who believes they have made the choices through a genuinely free act of will.

As long as Mick's mechanism for choosing between A and B is not visible to anyone else, Mick has gained the absolute minimum amount of knowledge. It's not quite zero, because he is now convinced that Shah knows the password, but beyond that fact, he has gained exactly zero. He can't even pass on his knowledge! There is a good reason that this is called a "zero-knowledge proof." Zero-knowledge proofs work only if Mick is free to choose his random sequence. And only Mick needs to know that his choices are free. It is not necessary for Char Lee or any other third party to know. This is fortunate, because as I will show shortly, no third party can tell whether Mick has free will without interacting with Mick. Observing him will never be enough, so no videotaping scheme will ever work if he indeed is free to choose his sequence.

I once went to an *avant garde* music concert at CNMAT, Berkeley's Center for New Music and Audio Technology, where the engineers had invented a new form of musical instrument that is played hands-free by a dancer. The dancer's motion in space is captured using a Microsoft Kinect, the same device used in video games, and a computer translates the dancer's motion into sound. But sitting in the audience, I could not tell that the dancer was not dancing to a recording. The music and the motion were certainly correlated, but they usually are with dance. Only the dancer could tell that it was he making the music. Interaction, by definition, is a first-person event. It cannot be externally observed. It has to be experienced. And only the dancer could experience the coupling of his motion with the sound.

Human cognition seems to have this same property; certain aspects cannot be externally observed. If cognition depends fundamentally on interaction, as postulated by proponents of embodied cognition (chapter 7), then we will never know whether machines possess it unless we become those machines. We have to gain the first-person perspective. Uploading our soul to a computer may be the only way to tell whether the computer can have a soul. And even if it can, and my uploading is successful, I will know, but I will not be able to convince you. The soul, whatever it is, may fundamentally be first-person knowledge, just like Mick Ali's knowledge that Shah Fi knows the password. That knowledge is obtained through first-person interaction, and that knowledge cannot be conveyed to any third person, at least not convincingly.

Even Mick's knowledge, however, is not certainty. Just as with Pearl's causal reasoning, some background assumptions are needed. Mick has to believe in his free will, and dismiss ideas like that Shah Fi is somehow manipulating his subconscious brain to make colluding choices. Ultimately, a little bit of trust is required to get past all the conspiracy theories. Once we open the door to trust, we have to admit that a third person may decide to trust Mick and assume that he is not colluding with Shah, in which case, despite Shah's wishes, her secret will be out. Maybe the lesson is to never collaborate with anyone trustworthy. Or maybe Shah is just being too paranoid.

The same level of trust is required for a third person to trust an RCT or a do-operator experiment. The first-person knowledge is simply not transportable, so if we don't admit some level of trust, even an RCT will not convince us. I will return to this in the next chapter when I examine what happens when powerful forces systematically seek to undermine trust.

WHO ARE MICK ALI, SHAH FI, AND CHAR LEE?

Zero-knowledge proofs were first developed by Shafi Goldwasser and Silvio Micali of MIT. Their first paper on this subject appeared in 1985 with co-author Charles Rackoff,[3] after earlier versions had been rejected three times by major conferences gated by peer review. As I hope you have come to appreciate, their idea is incredibly nuanced, so it's not surprising that expert peer reviewers would not get it. Goldwasser and Micali shared the 2012 Turing Award for this work.

Goldwasser and Micali have both made enormous contributions in cryptography and theory of computing besides zero-knowledge proofs. Both majored in math in college, Goldwasser at Carnegie Mellon University and Micali at Sapienza University in Rome, and then converged at UC Berkeley for their PhDs. At Berkeley, they wrote together one of the most influential papers in computer science, entitled "Probabilistic Encryption."[4] This paper defined mathematically, for the first time, what is a secret. They both converged again at MIT as faculty members, and it was there, in the lively computer science theory group, that they developed the idea of zero-knowledge proofs, although as a Berkeley guy myself, I have to believe that the ideas germinated at Berkeley.[5]

Zero-knowledge proofs have obvious applications in authentication using passwords, but they are still rarely used. Most modern password-handling software leaks information that fundamentally does not need to be leaked.

A particularly intriguing potential application is in nuclear disarmament. A graduate student at Princeton University, Sébastien Philippe, led an experiment that showed that the concept of zero-knowledge proofs could be used to verify that two physical objects, say, two nuclear warheads, are identical, without revealing any information about the structure of these two physical objects.[6] This could be used, for example, to prove that an object that is about to be destroyed is a genuine bomb without revealing anything about how the object is built.

COMMUNICATING WHAT CANNOT BE COMMUNICATED

A password is something that Shah can convey to Mick if she chooses to do so, even over a noisy channel, assuming that the password can be encoded with a finite number of bits. What if Shah wants to convey something to Mick that cannot be encoded with a finite number of bits? Suppose, for example, that Shah wants to prove to Mick that she is conscious. It is plausible, as I have argued, that her consciousness contains more than a finite number of bits of information. In this case, she can still use a strategy like that of a zero-knowledge proof, but she will only convince Mick up to some level of confidence short of one hundred percent.

I will now make a connection that, to my knowledge, has never been made before, to an idea in computer science known as bisimulation. This concept shows that it is possible to have two systems, say one that is conscious and one that is not, that appear to be identical if you merely observe them, but that can be distinguished if you interact with them. Moreover, as you interact with them, if you can only interact through the inputs and outputs of the system (you cannot peer into their "souls"), then you cannot achieve one hundred percent confidence, but you *can* get arbitrarily close to one hundred percent. To actually achieve one hundred percent confidence requires information about the internal state structure of a system, meaning what states it can be in and what states can follow from any given state.

If the state structure required to achieve cognition is transfinite (i.e., it cannot be encoded in a finite number of bits), then it cannot be realized digitally. Moreover, if cognition is a *learned* state structure, and if learning inevitably results in inscrutable machines (chapter 6), then even if humans create an AI that is conscious, we will never have one hundred percent confidence that we have done so.

Bisimulation is a formal, mathematical concept, and we have no mathematical model of consciousness, so everything I have just said about consciousness is really wild conjecture. However, I have argued that there is an aspect of free will that we *can* model formally, specifically, *when* the resolution of alternatives occurs. In a purely deterministic world, the resolution between alternatives occurs at the Big Bang, whereas in a nondeterministic world, it can occur later. The difference between these two situations is easy to model formally, and it turns out that this difference is not observable without interaction.

This does not address *why* or *how* the resolution between alternatives gets made, but it is plausible that in a human form of free will, these require a transfinite state structure. If free will arises from a tight interplay between randomness and control, as with Picasso's brain using his brush to render the dripping paint, then a transfinite state structure is a natural one.[7] The randomness of paint dripping is a continuous randomness, more like riding down a bumpy road than like flipping a coin. And the control of a brush to render paint is a continuous process, not the step-by-step algorithmic and discrete behavior of a Turing machine. If the continuousness of these processes is an essential feature of the system, then a transfinite state structure is also essential.

The concept of bisimulation arises in computer science in the context of the theory of automata. It may seem odd to apply a theory of "automata" to a question of free will because the common meaning of the word "automaton" precludes free will. But it turns out that the mathematical models that computer scientists call "automata" do not preclude free will. So with apologies, I ask you to tolerate the lexical dissonance and open your mind to what an automaton actually is to a computer scientist. Most of the credit for the concept goes to the English computer scientist Robin Milner (1934–2010). Milner, who spent most of his career at the University of Edinburgh and later at Cambridge, won the 1991 Turing Award for his contributions to machine-assisted proof construction, to type systems for programming languages, and to the theory of concurrent systems. His work on automata theory does not get mentioned in this citation, but to me, it was equally momentous. Let me illustrate the key ideas with a small example.

A TINY UNIVERSE

To understand the consequences of Milner's model, let us consider a tiny universe with just one being and nothing else. This being comes into existence at an event that we will call *bang*, and after it exists, it can do one of two possible things, *tick* or *tock*. The "laws of physics" of this universe are very simple: *bang* occurs exactly once followed by exactly one occurrence of either *tick* or *tock*, but not both. Nothing else is possible or even imaginable. This is the sum total of the laws of physics for this universe. This would be a very boring and lonely universe to live in.

In this universe, we do not need any detailed model of time. The life of the universe is either the sequence (*bang*, *tick*) or the sequence (*bang*, *tock*), and the only notion of time that is needed is that during the life of the universe, either no events have occurred, one event has occurred, or two events have occurred. We have no need for any measure of time between events nor any notion of the events occurring at some particular point in a measurable time continuum.

The question we want to focus on now is *when* the determination between *tick* and *tock* occurs. Figure 12.2 depicts a being named *Pablo* that makes the selection between *tick* and *tock* as late as possible. Figure 12.3 depicts an alternative being named *Edward* that makes the selection as early as possible, upon the occurrence of *bang*. *Edward* is the less creative

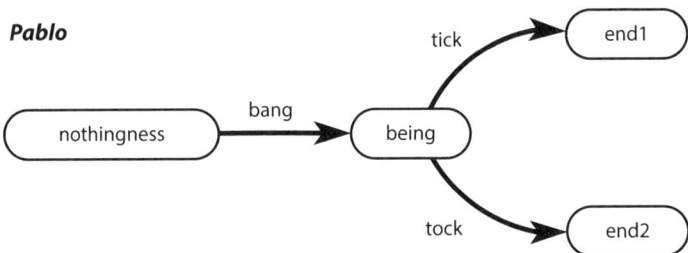

12.2   A being named Pablo in a tiny universe where a choice is made late.

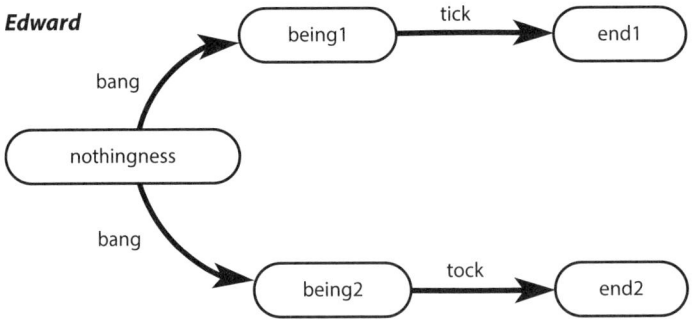

12.3   A being named Edward in a tiny universe where a choice is made early.

being because his entire future is determined when he comes into existence. *Pablo*, on the other hand, is more creative and potentially surprising. After he comes into existence, both *tick* and *tock* are still possible.

Both *Pablo* and *Edward* are compatible with the laws of physics of this tiny universe, as stated so far. But does this universe admit free will? Is it possible for *Pablo* or *Edward* to *choose* between *tick* and *tock*?

Edward lives in a universe where the determination of alternatives occurs at the time of the "Little Bang" (it is, after all, a tiny universe). *Pablo* lives in a universe where determination can occur after the Little Bang. In this sense, *Pablo*'s universe is nondeterministic and *Edward*'s is deterministic. Since *Edward* comes into existence at the time of the Little Bang, it is evident that *Edward* cannot possibly choose between *tick* and *tock*. An entity has to exist before it can take any action, such as making a choice. *Pablo*, on the other hand, exists at the time of the determination. It is possible, therefore, for some mechanism that we might call "free will" to be involved in the determination. This is arguably the smallest possible universe where we can make a distinction between free will and no free will.

On these diagrams, the bubbles represent states of the beings in the universe, and the arrows represent transitions between states. Such models are called "automata" by computer scientists (or, alternatively, "discrete transition systems"). Each universe starts in the *nothingness* state and then transitions to a *being* state. These states are given suggestive names, but the names are technically arbitrary. The *being* state name suggests that the being has come into existence, but neither *tick* nor *tock* has yet occurred.

In these models, when a state has two arrows coming out of it, this means that either arrow can be taken. A computer scientist would call *both* of these models "nondeterministic" because each has a state where there are two possible next states. But we are using *Edward* to represent a being in a deterministic universe that can come into existence in two distinct ways, and the determination occurs at the Little Bang, as the being comes into existence.

To a passive outside observer, an observer that only sees the *bang*, *tick*, and *tock* events, *Pablo* and *Edward* are indistinguishable. They both produce either (*bang*, *tick*) or (*bang*, *tock*). Such an outside observer has no way to construct a falsifiable hypothesis that the being has the structure of *Pablo* and not *Edward*. Hence, according to the philosophy of science due to Karl Popper, such a hypothesis is not "scientific."[8] An outside observer may "predict" *tick* after seeing *bang*, but if the observer sees *tock* instead, she still has no way of knowing whether the being is *Pablo* or *Edward*. The hypothesis that humans have free will may similarly not be scientific if it must be based on the observations of an outside observer. We will see, however, that an observer that can *interact* with the automata *can* tell the difference, at least up to some level of confidence short of one hundred percent.

In automata theory, *Pablo* and *Edward* are said to be "language equivalent" because they are both capable of producing the same "sentences" (*bang*, *tick*) and (*bang*, *tock*). But even though they are language equivalent, they have significant differences. How can we characterize those differences?

It would be easy here to fall into a trap and use our human understanding of the diagrams to explain their differences. *Pablo* makes decisions later than *Edward*, obviously. But what if we can't leverage such a higher-level function as human understanding? A more disciplined approach is to extend our universe in the smallest way possible to include *within the universe itself* the observer that must tell the difference between *Pablo* and *Edward*. If we succeed in doing that, then we can be sure that we haven't relied on some magical metaphysics.

A SLIGHTLY LESS TINY UNIVERSE

In this section, I will show that *Pablo* can model *Edward* but not vice versa. We will augment our universe so that it has two concurrent beings in it, one of which observes and "models" the other. This universe is less lonely because there are two beings, though I have to admit it will still be pretty boring. Each being is structured like either *Pablo* or *Edward*. We can now define what it means for one being to "model" another using a concept that Milner called "simulation." We can put the observer and observed beings into the same (slightly augmented) universe, and we have constructed what is perhaps the smallest possible universe capable of modeling.

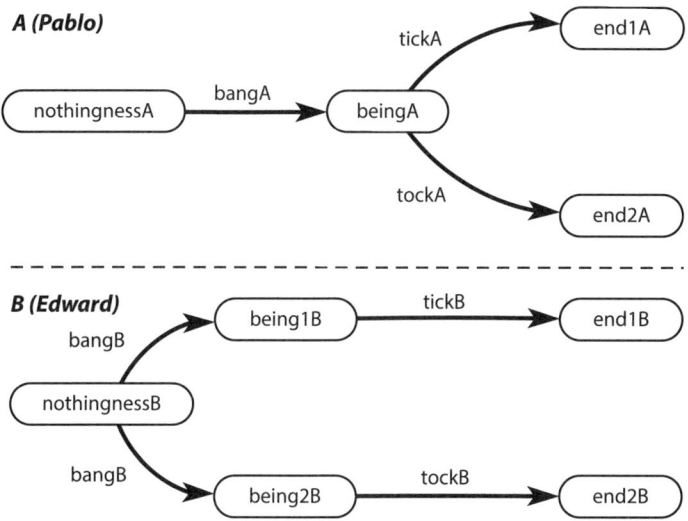

12.4 A small universe with two concurrent beings.

Figure 12.4 shows a universe with two beings, *Pablo* and *Edward*, that exist at the same time in the same universe. A dashed line between them is used to suggest *concurrent* composition, which means that both beings are active simultaneously.[9]

Let us give the diagram in figure 12.4 a very specific meaning (a *semantics*). We will assume that the beings *Pablo* and *Edward* take transitions simultaneously. They both begin in their *nothingness* states, so the initial state of the universe is the pair (*nothingnessA, nothingnessB*). At the time of the Little Bang, the two automata simultaneously transition and both beings come into existence. *Pablo* goes to *beingA* and *Edward* goes to either *being1B* or *being2B*. The resulting state of the universe is now either the pair (*beingA, being1B*) or the pair (*beingA, being2B*). In the next transition, there are four possible event patterns that can occur, (*tickA, tickB*), (*tockA, tockB*), (*tockA, tickB*), or (*tickA, tockB*). There are similarly four possible ending states for this universe.[10]

Now that we have two beings, we can begin to ask questions that involve modeling. Specifically, suppose that one of the two beings, say *Pablo*, is modeling the other, say *Edward*. What do we mean by "modeling"? We can give this loose term a specific formal meaning. Specifically, what we mean by "*Pablo* models *Edward*" is that *Pablo* attempts to mimic the moves that *Edward* makes. Specifically, whenever *Edward* produces *bangB*, *Pablo* matches it by producing *bangA*, whenever *Edward* emits *tickB*, *Pablo* emits *tickA*, and whenever *Edward* produces *tockB*, *Pablo* produces *tockA*. That's all. Modeling is imitating.

Now for the key insight. The universe in figure 12.4 is asymmetric. *Pablo* can model *Edward* but not vice versa. Since transitions occur simultaneously in the two beings, by the time of the second transition, *Edward* will have already had determined which of events *tickB* or *tockB* he will generate, but not *Pablo*. *Edward* therefore is incapable of matching every possible move of *Pablo*. On the other hand, *Pablo* is capable of matching any move *Edward* can make.

In automata theory, the ability of one being *Pablo* to model another *Edward* is called "simulation." Formally, *Pablo* "simulates" *Edward* if whatever moves *Edward* makes, *Pablo* will be able to match the events *Edward* produces and move to a state that will enable it to continue to match the moves of *Edward*. Note that when we say "*Pablo* simulates *Edward*," we are describing a scenario where *Edward* is free to make whatever moves its automaton allows, whereas *Pablo* will be constrained to make moves that match *Edward*'s events. *Pablo* is a passive observer of *Edward* in that it has no effect on the moves that *Edward* makes.

When a being *Pablo* simulates a being *Edward*, this fact can be represented by a "simulation relation," a set of pairs of states of *Edward* and *Pablo*. By convention, in this pairing, the states of the unconstrained being *Edward* are listed before those of the being *Pablo* doing the modeling. So for the universe in figure 12.4, the statement "*Pablo* simulates *Edward*" is represented by the following pairing of states:

Edward	Pablo
nothingnessB	*nothingnessA*
being1B	*beingA*
being2B	*beingA*
end1B	*end1A*
end2B	*end2A*

These pairings of states are a subset of the possible pairings that could emerge if *Pablo* were making no attempt to model *Edward*. But when such modeling is being done, these pairings are the only possible states of the universe.

Suppose we turn the tables and try to make *Edward* simulate *Pablo*. In this case, we will fail to come up with a simulation relation. Once *Pablo* is in state *beingA*, *Edward* will be in either *being1B* or *being2B*, and in either case, *Pablo* will be able to make a move that *Edward* cannot match. Now, in this augmented universe with two beings, the hypothesis that *Edward* is a model for *Pablo* is falsifiable by experiment. An "experiment" is simply a run of the universe through its two state transitions. Any particular experiment may not falsify the hypothesis, but the possibility of falsification is there. If the choices of *Pablo* are genuinely free, and if we are able to observe multiple runs of this universe, then very likely the hypothesis will eventually be falsified.

The construction of the simulation relation is a shortcut that makes repeated experiments unnecessary. In effect, the simulation relation represents all possible experiments all at once. If we can construct a simulation relation showing that *Pablo* simulates *Edward*, then we have proven that all experiments will support the hypothesis that *Pablo* is a model for *Edward*. But notice that in order to construct the simulation relation, we need to know the state structure of the two automata. We have to peer into their "souls" rather than just observe their behavior.

ASYMMETRIES

Pablo makes decisions as late as possible, whereas *Edward* makes decisions much earlier. *Pablo* lives in a world that may have free will, whereas even a compatibilist philosopher would have trouble asserting that *Edward* has free will, since this universe is too simple for any common sense notion of free will to exist within it.

We could construct a tiny universe like this with just two beings each with the structure of *Edward*. Now, either being can simulate the other, but interestingly, the observer *Edward* cannot model the observed *Edward* if it can only see the events it produces, the *bang*, *tick*, and *tock*. It has to be able to "see" what state the observed *Edward* moves into. In contrast, when *Pablo* models *Edward*, he can match the moves of *Edward without* seeing what state *Edward* is in. The events produced by *Edward* are enough. If we think of the output events *bang*, *tick*, and *tock* as "measurements," then modeling based on measurements is not really possible in a deterministic universe. You have to see more than measurements, peer into the soul of the system you are observing, and see exactly what state it is in.[11]

If our real universe is deterministic, then every model we construct must also be deterministic. It must have a structure like *Edward*, not like *Pablo*. To construct any model in such a deterministic universe, the model has to be clairvoyant. It has to "know" the entire future of whatever it is modeling, and it has to "know" this at the time of its creation.

The fact that we humans have a strong notion of free will suggests that we are able to model an automaton like *Pablo*. In fact, I have just done that in this book! If the universe in which this book exists is deterministic, then I don't see how I could have just done that. Any explanation I can come up with is far more complex than the much simpler explanation that the universe we live in is not deterministic.

BISIMULATION

Since our understanding of physics is mostly but not completely deterministic, then our model of the universe should be capable of determining outcomes both early and late. Some actions will follow deterministically from their preconditions, and others, like Norton's ball rolling off the hill, will happen nondeterministically. The automaton *Eduardo* in figure 12.5 models such situations by mixing features from *Edward* and *Pablo*. It has three transitions out of its *nothingness* state, two of which lead to deterministic consequences, and one of which postpones decisions until later.[12] It is easy to verify that *Pablo* simulates *Eduardo* and *Eduardo* simulates *Pablo*. Does this mean that we don't really need the extra complications in *Eduardo* that allow both deterministic and nondeterministic trajectories? Perhaps the simpler model *Pablo*, with only nondeterministic trajectories, is sufficient. This seems unlikely, however, because deterministic models in physics have historically proven extremely useful.

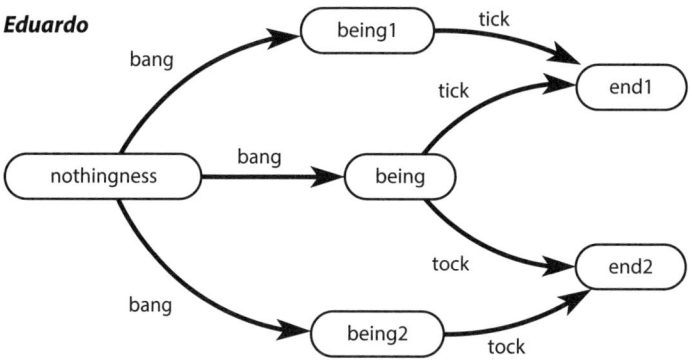

12.5 A tiny universe Eduardo that determines outcomes early or late.

In 1980, David Park, a computer scientist from the University of Warwick in England, came to Edinburgh for a sabbatical and moved into the top floor of Robin Milner's house. One day, Park came down to breakfast carrying Milner's 1980 book on concurrency theory and told Milner "there is something wrong." Park had found a gap in Milner's prior notion of simulation. He noticed that even if two beings simulate each other, they can exhibit significant differences in behavior when they interact. Milner's prior notion of simulation was unable to distinguish *Eduardo* from *Pablo*, but Park noticed that the differences are significant.

Milner and Park together came up with a stronger notion of modeling that they decided to call "bisimulation."[13] Milner then fully developed and popularized the idea. He showed that the difference between *Pablo* and *Eduardo* becomes evident only if *Pablo* and *Eduardo* can *interact* with one another. It is not enough to just observe each other, as he had done previously with his simulation relations. Interaction is more powerful than observation again!

In Milner's simulation game, the observed automaton is always unconstrained, and the observing automaton strives to match the behavior of the observed one. But suppose that the relationship between observer and observed is more interactive, more symmetric, where a dialog is possible. In one turn, maybe *Eduardo* is unconstrained and *Pablo* tries to match, and in the next turn *Pablo* is unconstrained and *Eduardo* tries to match. This better represents dialogs or any bidirectional interaction in our own physical universe.

In this variant of the game, the difference between *Pablo* and *Eduardo* becomes evident. In Milner and Park's terminology, these two automata are not *bisimilar*. Suppose that in the first turn, *Eduardo* is unconstrained and transitions to *being1*. Suppose then that in the second turn, when *Eduardo* is in *being1* and *Pablo* is in *being*, then *Pablo* moves first and chooses to move to *end2*, emitting *tock*. *Eduardo* will be unable to match it because he can only emit *tick*.

No such problem occurs if the same automaton moves first on each turn. The difference between *Pablo* and *Eduardo* is evident only through the symmetry of an interactive dialog. A one-directional monologue is not enough. This suggests that a universe in which components can observe *and act*, such that the action affects the observation, is richer than one that allows only observation.

Formally, a bisimulation relation, like a simulation relation, is a collection of pairs of states. To prove that there is no bisimulation relation between *Eduardo* and *Pablo*, we must show that no pairing of states ensures that whichever automaton moves first in each turn, the other will be able to match it. Such a proof, however, requires knowing the internal state structure of the automata. Without that knowledge, the best we can do is conduct repeated experiments, as Shah and Mick do with their cave.

SHAH AND MICK BISIMULATE

Shah Fi and Mick Ali's interaction in figure 12.1 can be modeled using the same sort of automata that Milner used, as shown in figure 12.6.[14] The model for Mick Ali is shown at the top. It shows that Mick enters the cave in the first time instant, then nondeterministically calls out *A* or *B*, ending in one of two possible states, *endA* or *endB*. The second model shows Shah Fi under the assumption that she does not know the password. She also enters the cave in the first instant, but nondeterministically goes to one of two locations, *insideA* or *insideB*. Once she is in one of these locations, she has no choice but to come out the same way she went in.

Notice that these two automata are identical in structure to those for *Pablo* in figure 12.2 and *Edward* in 12.3. As before, it is easy to verify that

Mick Ali

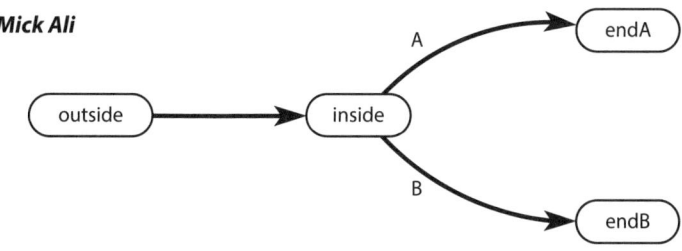

Shah Fi (Without Password)

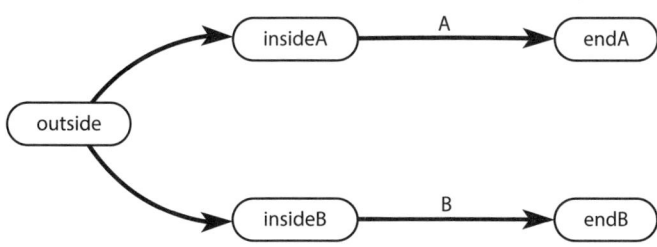

Shah Fi (With Password)

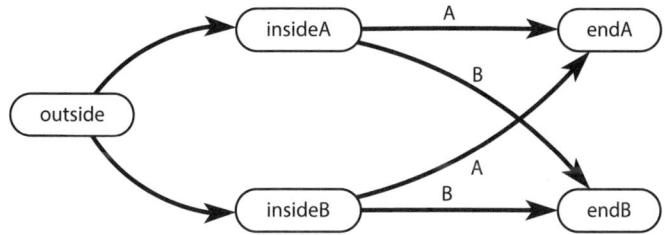

12.6 Automata models of Mick Ali and Shah Fi, with and without the password.

Mick simulates Shah but not vice versa. Shah is unable to make some of the moves that Mick may demand. The fact that Mick simulates Shah is what makes it possible for Mick to collude with Shah. Mick can match the decisions Shah has already made. Equivalently, Shah can anticipate whether Mick will call out *A* or *B*.

At the bottom of figure 12.6 is an automaton modeling Shah Fi under the assumption that she *does* know the password. In this case, it is easy to verify that Shah is bisimilar to Mick. They can perfectly match each other's moves regardless of who moves first at each time instant.

What really does bisimulation mean in this case? The two automata, Mick Ali and Shah Fi (with password), have different structure, but they are fundamentally indistinguishable. Mick's automaton represents what he demands from someone who knows the password. Shah's automaton represents the capabilities she acquires by knowing the password. The fact that these two automata are bisimilar shows conclusively what Mick is able to conclude with repeated experiments, that Shah knows the password. Hence, the repeated experiments provide evidence of bisimilarity that does not require knowing the detailed structure of the automata. But that evidence never gives one hundred percent certainty. To get to one hundred percent requires knowing the state structure.

Suppose instead that Shah guesses the password. This can be represented by the automaton in figure 12.7. This automaton has the same structure as that in figure 12.5. Here, if Shah correctly guesses the password, she is able to fool Mick no matter how many times they perform the experiment (assuming that the password does not change). This gives Shah's automaton the ability to simulate Mick's automaton. But as Milner and Parks showed, her automaton is still not fundamentally equivalent to Mick's. The possibility of guessing incorrectly remains. The lack of a bisimulation relation reveals the mismatch, albeit only if we know the structure of the automata. If we know that Shah's automaton has the structure shown in figure 12.7, then we know that she does not know the password, even if the possibility of a lucky guess remains.

Shah Fi (Guessing the Password)

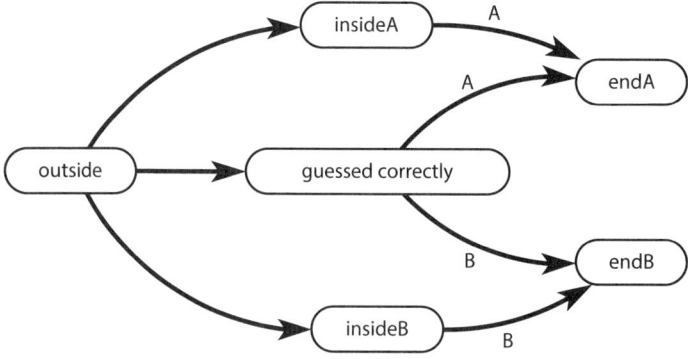

12.7   Automaton model of Shah Fi where she guesses the password.

HUMANITY REQUIRES INTERACTION

Interacting components can observe *and be observed* and can affect and *be affected.* Such interaction can accomplish things that are not possible with observation alone. The implications of this are profound. It reinforces Milner's observation that machines that look identical to an observer are not identical if you can interact with them. It reinforces Goldwasser and Micali's observation that interaction can do things that are not possible without interaction. It reinforces Judea Pearl's observation that reasoning about causality requires interaction (chapter 11). It reinforces the hypothesis of embodied cognition (chapter 7). If our sense of self depends on bidirectional interaction, the kind of dialog of Milner's model, where either party can observe or be observed, then our sense of self cannot be separated from our social interactions. Our minds cannot exist as an observer of the universe alone. And indeed, our interaction with the world around us has this bidirectional character. Sometimes we react to stimulus in ways that affect those around us, and sometimes we produce stimulus and watch the reactions of those around us. Such dialog seems to be an essential part of being human and may even form the foundations for language and possibly even thought.[15]

Moreover, such dialog has deep roots in physics. Quantum physics has taught us that no observation of a physical system is possible without disrupting the system in some way. In fact, quantum physics has real problems with any attempt to separate the observer from the observed. The observed automaton necessarily observes the observer. Passive observation in the form of unidirectional simulation is impossible in our natural universe. This suggests that simulation relations alone are not a reasonable model of modeling (a "metamodel," if you will permit me). Bisimulation is a better choice.

So, does it matter whether machines have a sense of self, the ability to engage in an interaction as a first-person participant? These results suggest that as machines increasingly engage in interaction, *acting* in the physical world rather than just observing it, they acquire the mechanisms to develop such a sense of self. But these results also suggest that we will never know for sure whether they have developed a sense of self. They will increasingly seem to have, and this will, no doubt, affect how we interact with them.

Although we get some modest insights from these tiny universes, to go further, we have to examine *why* and *how* the determination between alternatives is made. Looking only at *when* gives us some insights but not enough. This is particularly important as we cede more control over our lives to machines. They will be selecting among alternatives. How can we make sure that they do so on our behalf? This is the topic of the next chapter.

13

PATHOLOGIES

WHAT COULD POSSIBLY GO WRONG?

In 1951, the BBC Home Service started a radio program called *The '51 Society*, where a guest luminary would make a presentation and a panel discussion would follow. Alan Turing was invited three times. The first time he gave a presentation entitled "Intelligent Machinery, A Heretical Theory." In this lecture, he conjectured that,

machines can be constructed which will simulate the behaviour of the human mind very closely. They will make mistakes at times, and at times they may make new and very interesting statements, and on the whole the output of them will be worth attention to the same sort of extent as the output of a human mind....

Let us now assume, for the sake of argument, that these machines are a genuine possibility, and look at the consequences of constructing them. To do so would of course meet with great opposition, unless we have advanced greatly in religious toleration from the days of Galileo. There would be great opposition from the intellectuals who were afraid of being put out of a job. It is probable though that the intellectuals would be mistaken about this. There would be plenty to do, trying to understand what the machines were trying to say, i.e. in trying to keep one's intelligence up to the standard set by the machines, for it seems probable that once the machine thinking method had started, it would not take long to outstrip our feeble powers. There would be no question of the machines dying, and they would be able to converse with each other to sharpen their wits. At some stage therefore we should have to expect the machines to take control, in the way that is mentioned in Samuel Butler's "Erewhon."[1]

Erewhon, a novel by Samuel Butler first published anonymously in 1872, is perhaps the first serious reference to a society where machines have taken control. The novel features a fictional country, named "nowhere" spelled backward (almost, since the "w" and "h" are transposed). In *Erewhon*, the machines become banned because of the possibility that they may develop sufficiently by Darwinian selection to completely take over. Some took this as a joke, as if Butler was ridiculing Darwin, prompting Butler to write in the preface to his second edition,

> I regret that reviewers have in some cases been inclined to treat the chapters on Machines as an attempt to reduce Mr. Darwin's theory to an absurdity. Nothing could be further from my intention, and few things would be more distasteful to me than any attempt to laugh at Mr. Darwin.[2]

In his radio presentation, Turing confronted quite a number of risks, including that the machines will make mistakes; will clash with human spiritual beliefs; will "outstrip our feeble powers"; will become immortal; will conspire to improve themselves; will take control; and perhaps, through his reference to *Erewhon*, will evolve in a Darwinian way, beyond our control. He also anticipated generative adversarial networks (see chapter 4), saying that the machines will "converse with each other to sharpen their wits."

In his bestselling 2014 book, *Superintelligence*, Nick Bostrom, founding director of the Oxford Future of Humanity Institute, talks about an "intelligence explosion" potentially resulting in an "existential catastrophe" for humans. His premise is that when an AI learns to improve itself, an unstable positive feedback loop results (see chapter 5). Once the machine can make itself better, a runaway effect could occur where it will very rapidly outstrip the capabilities of humans and even the ability of humans to understand what is going on. Such a runaway superintelligence could be like a cancer to humans. Just like cancer, it could kill us, but also like cancer, we will do everything in our power to prevent that, once we recognize the cancer. Bostrom is not sanguine about us recognizing it in time.

Max Tegmark, featured in chapter 8 for his belief in teleportation, in his book *Life 3.0*, indulges in similar speculations. He argues that the real turning point is when computer software is autonomously designing and improving the hardware on which it runs. Computers already play a huge role in designing hardware, but today, human designers are still essential to the process. When they become inessential, Tegmark argues, a runaway effect will occur, and some scenarios will sideline humans altogether.

In this chapter, I would like to focus on the various ways that a world with digital machines can go wrong. I will begin with the fragility of machines themselves, but then focus on the ways in which humans, ensnared in a deep codependence, could be hurt. I will argue that these problems are not an *existential* threat, per se, unless what you mean by "continued human existence" is that things should not change from what they are now. They *will* change, but humans will continue to be part of the emerging obligate symbiosis, as I will argue in the next and final chapter. In this chapter, I suggest that the problems that will inevitably emerge be viewed as pathologies and be treated, like a disease, rather than being viewed as existential threats with which we should go to war. We may face runaway epidemics of screen addiction, unemployment, or technology-induced depression, but there will also be more efficient use of resources, more cross-cultural interaction, and richer, more dynamic intellectual lives, as we struggle to keep our own intelligence "up to the standard set by the machines."

THE DEATH OF A MACHINE

In chapter 8, I pointed out that a digital, algorithmic machine, such as a Turing machine, can, in principle, be immortal. It can be backed up and later restored, it can be migrated to new hardware without losing any of its essential features, and it can transported over arbitrary distances at the speed of light. This would seem to give machines an aura of invincibility, like the Roman gods. But they are more fragile than it might seem. Perfect copying of digital bits is only *theoretically* possible, as shown by Claude Shannon. In practice, there is always at least a small probability of error, and even a gargantuan machine can be brought to its knees by a single erroneous bit. Moreover, most interesting machines are not purely digital, but rather exist embodied, with a physical-world component that is essential to their being. They are not completely independent of the physical stuff of which they are made, except insofar as they are mathematical abstractions interacting only with our information world.

My friend Malcolm McCullough, a scholar who has written widely about how digital technologies have defined for humans new notions of space and craft, reported to me a dead machine in his very own kitchen:

The control circuit board for the range hood has shorted and smoked and is just old enough (12 years) that a replacement cannot be had, and the service tech guy, who won't even think about swapping in a more generic controller, has declared the entire hood dead. $1K worth of pretty stainless steel and fans gone for want of a couple switches and relays. To end so unambiguously and non-incrementally despite how many parts might still be working does somehow resemble life. One minute buzzing like a fly, another minute the same materials lying there, sans process, like a swatted fly.[3]

Swapping in a "more generic controller" is unlikely to work, like replacing the brain of an animal with one from another species. Unlike an AI running in a Google server farm, McCullough's hood was a distinctly embodied machine, an only partially digital being. Dr. Tech Guy was unable to restore it from backups, and therefore declared it dead at the scene.

CHIMPS WITH HUMAN BRAINS

There are other more sophisticated technological artifacts that are similarly mortal. Modern cars and airplanes include many digital parts and a great deal of software. But they too cannot be restored from backups when their subsystems fail. Even if it is their *digital* parts that fail, the whole organism may die.

I heard a story from an engineer who had worked on the 777, Boeing's first fly-by-wire airliner. ("Fly by wire" means that pilot controls are mediated by a computer.) The 777 first entered into service in 1995. According to this engineer, as of the early 1990s, Boeing expected this model of aircraft to be in production for perhaps fifty years, to 2045, and to be flying for another fifty years. The engineer told me that in the early 1990s, Boeing purchased a one-hundred-year supply of the microprocessors for the flight control systems so they could use the same microprocessors for the entire production run and maintenance of the aircraft.

Engineers at Boeing, Airbus, and several car companies have since validated this story. Replacing microprocessors with newer models simply does not work, like putting a human brain into a chimp. An Airbus engineer told me that they store the microprocessors in liquid nitrogen, hoping to extend their shelf life. Imagine if when you were born, the hospital sent you home with a freezer full of spare brains to be used to replace the one in your skull when it failed.

If the "cognitive" functions of the airplane reside in its software, then any hardware that correctly runs that software should be able to replicate those functions. Any modern microprocessor will "correctly" run any of the flight control programs in the 777. It will perform exactly the operations specified by the software with very high reliability. But that is not enough for the plane to fly. Evidently, the "cognitive" functions of the airplane are embodied, not resident solely in the software. As I pointed out in chapter 9, this is almost certainly true of humans as well, a message that will profoundly disappoint those who hope to upload their souls. Even if they include many digital and computational operations, the cognitive functions of humans are not just software running on a Turing machine.

When Boeing runs out of microprocessors, 777s will start dying of brain disease. In the meantime, however, brain transplants do work, thanks to the largely digital nature of those microprocessors and the cache of frozen spare parts. If McCullough had a spare controller for his hood in his freezer, Dr. Tech Guy would no doubt have been able to perform a successful "brain" transplant. But the "brain" of the hood is nowhere near as sophisticated as the control systems of even a bacteria. And despite its bulk, the cognitive sophistication of a 777 is much less than that of a fly, so perhaps it too relies on its relative simplicity for "brain" transplants to be successful.

AGING MACHINES

Despite being digital, even pure software systems, those with no significant embodiment, age. Banking software written in COBOL forty years ago may still be "alive," but its days are numbered. The US government is famous for keeping very old software alive well beyond its natural lifetime.[4] These programs will die eventually.

Any software system that grows, such as the astonishingly complex networked software that makes up the Amazon or Facebook colonies of services, runs the risk of getting crushed by its own complexity or developing a cancer that brings the whole system down. These systems are designed to be robust, and they tolerate even major amputations, like when Google pulled the plug in 2018 on their Google+ service. Perhaps this robustness will lead to immortality, but I suppose we have to wait a very long time before we can be sure. There are no programs running today that are older than about fifty years.

Turing computation is multiply realizable (see chapter 7). It can be perfectly replicated on different hardware and at different times. But an embodied machine, one that interacts with its physical environment, is more than a Turing computation, just as a human is more than a logic function realized by neuron firings, contrary to McCulloch and Pitts (see chapter 7). Turing computations observe inputs, perform step-by-step (algorithmic) operations, and produce outputs. These are purely observational operations, lacking feedback and incapable of reasoning about causality, forming a first-person self, exercising free will, and turning randomness into creativity (see chapters 11 and 12). Turing computation is a powerful tool, but it is not the universal machine that many people claim. Nevertheless, we can use it as a building block to build bigger, more interesting systems that go beyond Turing computation. Many of those systems, though perhaps not all, will be mortal.

MACHINE RIGHTS

Inevitably, when I talk about the possibility that technological artifacts can be viewed as living beings, the conversation turns to whether they should have rights. But being alive does not imply that you have rights. Lettuce and bacteria do not have rights in any society that I know of. (A possible exception is Jainism, but even Jains have to eat.) In fact, for much of human history, even humans did not have rights, and even today, rights are unevenly assigned. Yuval Noah Harari, who defined "dataism" as a quasi-religious philosophy (see chapter 3), reminds us that the sanctity of human life is a recent phenomenon.

We are constantly reminded that human life is the most sacred thing in the universe. Everybody says this: teachers in schools, politicians in parliaments, lawyers in courts, and actors on theatre stages. The Universal Declaration of Human Rights adopted by the United Nations after the Second World War—which is perhaps the closest thing we have to a global constitution—categorically states that "the right to life" is humanity's most fundamental value. Since death clearly violates this right, death is a crime against humanity, and we ought to wage total war against it.[5]

This universal "right to life," which is not really universal, evolved in our culture over a few hundred years, a blink of an eye in the grand scheme. What's to say that a principle of machine right to life will not similarly emerge?

Some machines will die at the hands of humans who kill them because they are harmful or offensive. Recall Tay, Microsoft's chatbot that learned from Twitter users to write vulgar and racist words. Microsoft executed Tay without even a jury trial. Recall the viruses and worms of chapter 2. Most computers run anti-virus software that seeks out and destroys such viruses. If machines are living, should we be worried about the ethics of these executions?

One way in which individual rights can emerge as a central philosophy is through a thought experiment that American political philosopher John Rawls called the "original position." In his celebrated 1971 book, *A Theory of Justice*, Rawls suggested that political decisions should be made from behind a "veil of ignorance," where the decision maker does not know "his place in society, his class position or social status," or "his fortune in the distribution of natural assets and abilities, his intelligence and strength, and the like."[6] The principle is that when you act to change a societal practice, you should do so under the assumption that you could find yourself in any role in the resulting society. If we include in that society machines, then it is not so farfetched that there will be movements toward granting them rights. Initially, the advocates are likely to be those who believe they will be able to upload their souls to computers (see chapter 9) because this makes it easy to imagine yourself in the role of a machine.

Rawls explicitly excluded from his theory nonhuman creatures because they do not participate in society as moral agents, where he defines society as "a cooperative venture for mutual advantage." But it is not hard to see how the same veil of ignorance could be used to reason about the rights of animals. Robert Garner, for example, extends rights to animals based on a presumption of their sentience.[7]

Since sentience is fundamentally a first-person property, such extension of rights must originate from imagining oneself inhabiting the body of the animal. In Jainism, an ancient Indian religion, the principle of *ahinsa* dictates avoiding harming *any* life form. Vegetarianism and other nonviolent practices follow from this principle. Ahinsa can be understood in conjunction with Jain notion of the karma, which posits that every soul reincarnates repeatedly in many life forms, including not just human, but also other animal forms. The resulting sentiment of kinship stems from imagining oneself as that ant and hence avoiding stepping on it.

If uploading ever becomes possible, then Jain-like reincarnation as a machine becomes a reality. Rawls's original position combined with uploading naturally leads to some level of machine rights. I have argued that uploading is impossible because of the Shannon channel capacity theorem (see chapter 8), so short of some new religion arising with a new theory of reincarnation as software, I do not see how to use the original position to argue for machine rights.

CAN WE KILL WHAT WE CREATE?

If we pull the plug on computers, we kill the digital beings they host, just as stopping the heart of an animal kills the animal. But the computers and their software are our own creation, aren't they? They would not exist were it not for us, so why should we hesitate to cause them to cease to exist? Of course, our children would not exist were it not for us, but we have no right to kill them.

Perhaps rights should only extend to members of our species as a matter of principle. There are few people, however, who would be unmoved by deliberate and unnecessary cruelty toward animals. People who are so unmoved are probably also unmoved by cruelty toward other humans, and we would likely classify them as sociopaths. Whether animals are our own creation or not is largely irrelevant. Modern cats and dogs have been genetically engineered by humans, but if anything, we are more likely, not less likely, to feel empathy toward them compared to animals that we have not had any hand in creating. Even nonvegetarians are repulsed by animal cruelty, showing a preference for free-range farming and quick slaughtering methods.

Will we ever reach a stage where empathy toward machines is natural and expected? We are already getting surprisingly close to this. In chapter 10, I talked about studies of humans behaving "abusively" toward robots. To even formulate such a study requires forming a notion of abuse, which in turn requires some measure of empathy. A can being kicked down the road is not being abused, but a robot being kicked by a kid is.

No digital machine today has anything like what we would call a sense of pain, humiliation, or embarrassment. Moreover, none today has anything like empathy for humans or other animals, so machines that are engaged in our society can probably be legitimately classified as sociopaths. HAL, the computer in *2001: A Space Odyssey*, exhibited distinctly sociopathic behavior.

I believe it is incorrect, however, to view and judge machines independently of their human symbionts. Microsoft, a company entirely dependent on software for its existence, created Tay, the foul-mouthed machine, but Tay was not held accountable for her actions. Microsoft was. When a human with a gun does something bad, we do not hold the gun responsible. The National Rifle Association in the United States lays *all* the accountability at the feet of the individual human, but more reasonable people understand that the human combined with a culture around guns is the real culprit. Microsoft was not entirely responsible for Tay's misbehavior either; part of the responsibility lies with the Twitter followers who taught her supervulgarity.

DISEASES

In chapter 1, I described how Rhonda, my wife, vehemently objected to my Amazon Echo and my Kubi. But she is no Luddite. When they first came out, she got a Fitbit, one of the first wearable devices that would track your steps and report back the distance you had walked each day.

Although she is not a fitness buff and does not have a gym membership, she believes strongly in "free exercise," walking when you could have driven and climbing the stairs when you could have taken the escalator. The Fitbit, however, pushed her over the edge. She started an internal competition to outwalk herself each day, never letting a day go by without at least five miles, and sometimes exceeding ten. Within a few months, she had developed plantar fasciitis, a painful condition in the feet that made walking extremely difficult. With the help of expensive shoes and a less self-competitive approach, she is back to doing a lot of walking, thankfully.

Rhonda is not alone. Many people have gotten obsessive about self-tracking, as if they were trying to create a digital mirror, an exoself living in the cloud, or a pixelated person.[8]

Technology affects our behavior, and our behavior affects our bodies. Several years ago, I personally suffered a severe bout of carpal tunnel syndrome that was triggered by intensive daily keyboard work on a textbook I was writing. I had to learn stretching exercises and better posture in order to peacefully coexist with my most important cognitive prosthesis, my computer. A former student of mine suffered a similar bout and started using speech recognition software to replace typing, only to develop pathologies in his larynx that, for a period of time, made it extremely difficult for him to speak. Both of us have recovered and continue to make extensive use of our computers, but we had to learn moderation.

The reality is that software changes our behavior, our bodies, and even our cognitive selves. Hardly a day goes by without news articles about Twitter addiction, cyber bullying, and connections between social media usage and depression and suicide in teens. The YouTube recommendation engine watches us, sucks us in, and eats our hours, disrupting our work and our lives. Evolution designed the human body, our culture, and our minds for success in a world we no longer live in.

Disrupted equilibrium is certainly not new to humanity. The evolutionary biologist Kevin Laland, who appeared in chapter 3 in the context of our discussion of IQ, states that:

The ideology of this society is that innovations solve problems, but that is only half of the story. Innovations construct new niches, just as organisms do, and every "solution" has the potential to generate many new "problems." When our ancestors first devised agriculture, they opened up a Pandora's box, and let loose the evil of the Anthropocene.[9]

That every solution generates new problems simply means that we need to keep working, not that we need to roll back the clock. Indeed, rolling back the clock could be disastrous.

WANT TO CRY?

Consider for a moment what would happen if today, as you sit there reading this, all the planet's computers were to be permanently turned off. If every computer were to suddenly die, all the money you could walk away with would be that stashed in your wallet right now. Go ahead: look and see how much money you would have. How long would that last you? All our phones would stop working. Most of the cars going by on the road would stop dead in their tracks. The lights would go off. Your heating system would stop working. Eventually, billions of humans would die from starvation or strife. Among the few technological artifacts that would continue to work would be guns.

We do occasionally see relatively minor forms of this Armageddon. On May 12, 2017, a massive worldwide malware attack called WannaCry disabled some 200,000 computer systems worldwide (see figure 13.1).

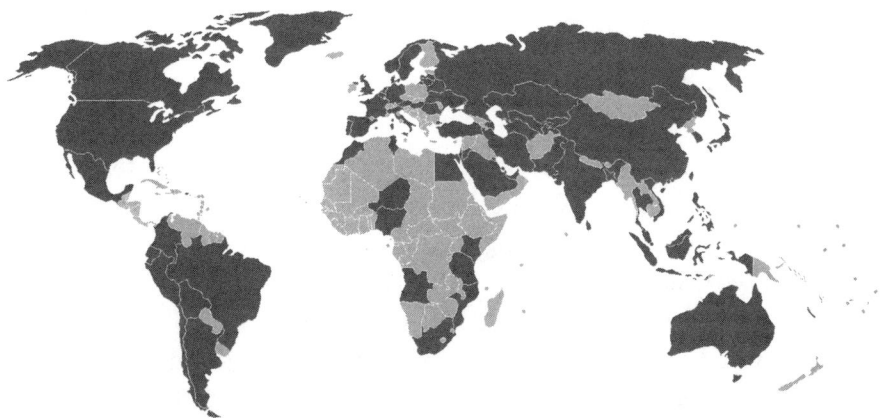

13.1 Countries initially affected by the WannaCry ransomware Internet worm. By User:Roke, CC BY-SA 3.0, via Wikimedia Commons.

Many of the hospitals of the National Health Service of England and Scotland had to turn away noncritical emergencies. Factories worldwide were forced to shut down, and the Spanish telephone company, FedEx, Indian police, and the German train system were all affected. This was not just an inconvenience that kept a few people from reading their email.

WannaCry took advantage of a vulnerability in Microsoft Windows systems that were not up to date. The malware program, classified as a worm (see chapter 2), broke into systems and encrypted their data using a secret key and then demanded ransom payments in Bitcoin to provide the decryption key. The price was $300 if the ransom was paid within three days and $600 if paid within seven days. Since Bitcoin payments are visible to anyone, it was easy to determine that as of June 2017, after the attack had subsided, over $130,000 had been paid, and nobody had reported getting the decryption key.

Curiously, the vulnerability in the Windows system that WannaCry exploited is believed to have been discovered by the US National Security Agency, who then wrote code that could exploit the vulnerability for use in its own offensive programs. This code is believed to have been stolen and then leaked by a group known as The Shadow Brokers, who have leaked several other exploits stolen from the National Security Agency.

In December 2017, the United States, United Kingdom, and Australia accused North Korea of being behind the attack.[10] Apparently, there was evidence in the worm files that the computers used to create the worm had fonts installed for the Korean Hangul writing system and were set to the Korean time zone. Also, the code had some similarities to code used in previous attacks linked to North Korea.

There have been many other disruptive attacks, including for example the 2016 attack on Dyn, a company providing a critical service needed to keep the Internet running. This attack hijacked millions of Internet-connected devices such as printers, baby monitors, and nanny cams, reprogramming them to flood Dyn's servers with useless requests, effectively preventing the servers from responding to legitimate requests. The attack seriously disrupted many Internet services, including Airbnb, the BBC, the Boston Globe, Comcast, Visa, and many others.

Some disruptions do not appear to be malicious. For example, on May 27, 2017, British Airways cancelled all their flights into and out of

Heathrow Airport, their primary hub, because of a computer malfunction. Thousands of passengers had their plans disrupted. Partner airlines were also affected, resulting in many more flight cancellations. More than seven hundred flights were cancelled over three days, and more than seventy-five thousand passengers were affected.

Like a bad cold, in each of these cases, it took a few days for things to get back to normal, and no lives were lost. In most of these cases, with the possible exception of the British Airways malfunction, humans were the primary agents both initiating the attack and building the "immune system" response. To counter the attacks, researchers study the code and find ways to prevent it from spreading, shut it down, or, in some cases, reverse its effects. Some industry pundits, such as the Austin, Texas, entrepreneur and author Amir Husain, argue that in the long run, only AIs will be able to counter these threats,[11] in part because it will be AIs, rather than North Koreans, designing and launching the attacks. We can envision AIs battling AIs, learning from each other and mutating to gain maximum advantage. The analogy with the coevolution of biological pathologies and immune systems is hard to ignore.

Our livelihood has come to depend on computers only recently, but it has always depended on other biological creatures. Consider what would happen to you if today, as you sit there reading this, all the bacteria in your body were to die. You may survive for a while, but you will be very sick. Biologists refer to our relationship with our gut bacteria as a mutualistic symbiosis, where both species benefit. Our relationship with machines may be becoming stronger, what biologists call an obligate symbiosis, where neither can live without the other. But just as our gut bacteria can turn nasty, technology can turn nasty.

Even when technology is operating normally, there will be inevitable costs to offset the benefits. Technology makes us smarter, at least collectively if not individually, but it also distracts us. It improves our health, but may threaten the mental health of our kids. It creates jobs, and it takes them away. Just like gut bacteria that give us cravings for unhealthy foods, technology is so intertwined with our very essence that its balance of positive and negative forces should be viewed as a question of health rather than as a competition.

DEEP FAKES

Nobody needs to remind us that the availability of information has exploded. At first, this seemed to me like it could only be a good thing. I thought, for example, that the Internet would spread enlightenment, democracy, and freedom, but it seems it is just as good at spreading hate. Stalwart democracies like the United Kingdom and the United States are voting themselves into factionalized partisanship based mainly on ignorance and dogma, both apparently fueled by floods of information. The Arab Spring, the 2010 uprisings against repressive regimes in North Africa and the Middle East, was driven by freer information flow but quickly chilled into an Arab Winter, and very little has improved.

There is no question that the flux of information flow has dramatically increased over the last few years, but the problem may be that the information ecosystem still involves humans, and our capacity to handle information is limited. The machines at Google and Baidu are as close as we've gotten on this material Earth to anything omniscient, but they can't share that omniscience with us because we simply can't handle it. Instead, they profile us and curate our information based on our prior behavior, something that can easily result in a flood of information that feeds our preconceived biases and starves our skepticism. We *think* we are acquiring more information, and indeed we are, but there is plenty of information out there to reinforce what we already believe, more than we can possibly absorb, so there is no need to feed us anything we might not like. As a result, we see vast amounts of information, but all of it within a "filter bubble," a curated subset, creating islands of disjoint truths, as discussed in chapter 3.

Information exchange between humans is the backbone of culture. Kevin Laland, an expert on animal behavior, in his book, *Darwin's Unfinished Symphony*, shows that a key tool in nature is the ability that many animals have to learn from each other by copying behavior. He shows that what we may call "culture" arises in many species, not just in humans. In the resulting societies, animals that master strategic copying fare better than those that don't.

Humans not only copy the behavior of those around us (we certainly do that), but we also learn from people whom we have never met and who may even no longer be alive. Laland points out that human learning is enormously magnified, compared to lower animals, by this ability to learn from people we have never met. This ability, in turn, is enabled by language, writing, teaching, and the cultural practices surrounding these. Laland argues that it is primarily this cultural magnification that distinguishes humans from other animals and enables a society comprising billions of individuals overwhelming the planet. Our computers dramatically enlarge this magnification.

For a society to work, it has to be built on shared information and shared beliefs. We all believe in the objectively meaningless pieces of paper that we call money, for example. The idea of money is a shared belief, and a society at our current scale could not possibly work without such shared beliefs. Most aspects of our lives depend on shared cultural values that are learned through our information exchange mechanisms.

Until recently, spreading information was relatively costly, requiring investment in printing presses, distribution channels, television and radio broadcasting rights and equipment, and so on. As a result, most sources of widespread information were institutional, and the institutions built reputations, good and bad, and business models around those reputations. As a society, we have developed a largely shared value system that trusts some sources of information over others.

Is trust really necessary? Perhaps there is some way to dispassionately, objectively evaluate information. We say, "seeing is believing," so perhaps if we can observe something with our own eyes, we can trust its veracity. For this to scale to large populations, however, we need photos and videos, since our eyes cannot be everywhere. But photos and videos have always misrepresented reality to some degree, and today, they can be convincing and yet completely fake.

In Fall of 2017 an anonymous Reddit user with username Deepfakes published software to make synthetic videos with one person's face substituted for another's. The resulting video even matches the facial expressions of the original. As a demonstration of the software, the user posted pornographic videos that appear to feature various Hollywood actresses. It is difficult to tell that the videos are fakes.

Many AI researchers have picked up on this theme. A team of Berkeley researchers led by Alexei Efros created software that can generate a video of any person making the dance moves of a professional dancer, for example.[12]

The Reddit deepfake software was built using generative adversarial networks, which pit neural nets against each other (see chapter 4). Some of the world's best AI researchers are now engaged in building AIs that can identify fake videos, escalating this adversarial arms race to a new level.[13]

There is even a DARPA program called Media Forensics, or MediFor for short, addressing the threat that synthetic media pose to national security. Although it is scary enough that fake but convincing videos can be created, an even scarier proposition is that anyone can cast doubt on even undoctored real videos. Consider that every politician will be able to denounce as fake any incriminating audio or video recording. Not only has the published word become suspect as a means for communication, but every publication medium we have has been undermined. Does this mean that we will no longer be able to learn from people we have never met?

INFORMATION APOCALYPSE

Learning from people we've never met means learning without interaction, and as we saw in chapters 11 and 12, some things cannot be accomplished without interaction. Recall the parable of Mick Ali and Shah Fi's cave. If Mick Ali is the teacher, he can teach us that Shah Fi has the password only if we trust him. If we suspect that Mick is colluding with Shah, then we will not learn. Mick becomes not a teacher but a propagandist, a conspirator in an effort to deceive us. Only by trusting Mick can we learn what he has learned. If we can't even be sure that Mick is a human and not a chatbot, are we likely to trust him?

When we receive information from a source we do not know personally, we are more likely to trust that information if it conforms closely with what we already know. Unfortunately, this means we may build trust in information sources that craft fakes to match our personal predilections. This happened repeatedly during the 2016 presidential election in the United States, where, for example, posts declaring that the Pope had endorsed Donald Trump were directed to people likely to believe this.

Technology has democratized publication and broadcasting to the point that any individual can reach every individual. The cost of distributing information globally has dropped to (essentially) zero, and the number of sources of information has exploded. No human can absorb even a tiny fraction of this information flux, and yet, we trust those humans to vote for presidents and to seal our national borders. Curating the information flow to individuals becomes essential, but if it is based on a principle of not feeding the individual anything they might not like or believe, we can only expect a disaster due to polarization.

Information flows have always been curated, but historically, this curation was not individualized. In the 1950s, there were three major television broadcasters in the United States, and the information they curated for us went to nearly everybody in the country. In such an environment, a common culture values *trusted* sources of information, and sources that violate that trust quickly lose currency. The mechanisms we are now using for information dissemination do not naturally cultivate any shared notion of a trusted source, and even our highest leaders systematically subvert what little common trust is left.

At the time of this writing, Donald Trump, the Republican Party, and both right- and left-wing extremists are barraging the American public with a message that the media and other authority figures in our culture (scientists, academics, and political leaders) cannot be trusted. Trump himself, the elected President of the United States, has made himself the paragon of a political leader that cannot be trusted. He lies and knows he is lying, undermining the authority of his office.

For me, the Trump phenomenon has been a revelation. In my view, the forty percent or so of Americans who continued to support him after his first three years in office should have been ashamed of themselves, but they weren't. I've deliberately expressed my opinion here in the most starkly partisan terms I can muster to make a point. My reality is clearly not the same reality as that of millions of other Americans. My background principles of decency, morality, and fairness make it impossible for me to understand the position of those people, a situation that I suspect is symmetric. How did we get here?

Harari, in *Sapiens: A Brief History of Humankind*, argues that the ability to construct societies that are more than tiny tribes depends on media communication. Trump's broad attack on societal trust in such communication is subversive. If he succeeds, he will throw us back to the tribal world of illiterate apes. The only information we will trust is that we directly obtain from the observed behavior of those around us. If your neighbor believes that earth is four thousand years old, you will too. If your neighbor believes that climate change is a hoax, you will too. Anything written about this by strangers, proclaimed by political leaders, or reported in *Science*, will be dismissed as not trustworthy.

Daniel Dennett, in his book, *From Bacteria to Bach and Back*, talks about how human communication augments the power of a fundamentally limited human brain:

Human brains have become equipped with add-ons, thinking tools by the thousands, that multiply our brains' cognitive powers by many orders of magnitude. Language, as we have seen, is the key invention, and it expands our individual cognitive powers by providing a medium for uniting them with all the cognitive powers of every clever human being who has ever thought. The smartest chimpanzee never gets to compare notes with other chimpanzees in her group, let alone the millions of chimpanzees who have gone before.[14]

Today, we have spectacularly more effective mechanisms for "comparing notes," but if we lose trust in the notes we compare, we lose this cognitive multiplier and become chimpanzees.

In *Darwin's Unfinished Symphony*, Laland points out that "teaching" as practiced by humans is absent in all other animals. Teaching, according to Laland, has no evident evolutionary purpose, since the conveyance of knowledge from the teacher to a stranger does nothing to help propagate the teacher's genes. Teaching is evolutionarily altruistic, and therefore inexplicable from the perspective of purely biological evolution. Today, teaching is under attack, in part because of widespread distrust of intellectual authority. Ironically, some of this distrust stems from a rejection of the scientific theory of evolution in favor or religious theories, but the attack on teaching may throw us back to the early days of evolution, where only biology, not knowledge, made progress.

Are we facing an information apocalypse?[15] According to Wikipedia, which ironically has been gaining trust in an era where everything else is losing it, the word "apocalypse," from ancient Greek, literally means "an uncovering." It is a disclosure of knowledge, a revelation. In Christian tradition, from the Apocalypse of St. John, the last book of the New Testament, the revelation which St. John receives is that of the ultimate victory of good over evil. It signifies the end of the present age. Today the term is commonly used for any prophetic revelation that leads to an end of time or to the end of the world. It may well be that the information flood unleashed by technology, which is swamping our feeble brains, spells the end of the present age. It was much easier to agree when there was less information flowing and most sources of information could be trusted.

The American historian of science James Gleick, in his book, *The Information*, points out that the very notion of "history" is inextricably tied to media.[16] Without a mechanism for communicating beyond our personal encounters, there can be no history. He goes further to claim that the very act of "thinking" and our notions of logic and reason are also tied to writing, as is even the notion of consciousness. When writing and other media lose their veracity and authority, when we lose faith in them, we lose history, thought, reason, and consciousness. This sounds like an apocalypse to me, but not the one of St. John. It would be a triumph of evil over good.

Is this apocalypse likely to occur? This depends on how we, as a society, react to the fact that far more information is readily available than any individual can possibly absorb and the fact that any medium can be faked. Each individual becomes vulnerable to buffeting by curated data, curated by buffoons, AIs, nefarious politicos, and corporations, to shape our consciousness. Only by building trust can we prevent this apocalypse.

TAKE BACK THE WEB

One cause for the filter bubbles that the Internet creates for us is that we have freely given up privacy, revealing the most intimate details of our lives to the AIs at Facebook, Google, and Amazon, and in China, at Tencent, Baidu, Alibaba, and the Chinese government. Harari, in *Homo Deus*, says these AIs are coming to know us better than we know ourselves. And these AIs, designed to maximize advertising revenue, exploit that information to feed us only what we most want. As Stuart Russell points out (see chapter 3), the information the AIs feed to you changes *you* to make you more effective toward achieving the goal of the AI, to maximize revenue.[17]

Tim Berners-Lee created the World Wide Web three decades ago by developing a "hypertext markup language" (HTML). He has been recognized for his enormous impact on the world: he received a Turing Award, was named one of the twentieth century's most important figures by *Time*, and he was knighted by the Queen of England. But he has recently expressed profound and painful disappointment in what the web has become. He laments the simultaneous massive monopolization phenomenon, where a few huge corporations have taken control of all the data.

In an article in *Vanity Fair*, Katrina Brooker quotes Berners-Lee, saying, "We demonstrated that the Web had failed instead of served humanity, as it was supposed to have done, and failed in many places." *Vanity Fair* summarizes his position as follows:

The power of the Web wasn't taken or stolen. We, collectively, by the billions, gave it away with every signed user agreement and intimate moment shared with technology. Facebook, Google, and Amazon now monopolize almost everything that happens online, from what we buy to the news we read to who we like. Along with a handful of powerful government agencies, they are able to monitor, manipulate, and spy in once unimaginable ways.[18]

Targeted advertising has created a surveillance society. And the situation seems to be getting worse. Kai-Fu Lee, the former of director of Google's China operation, who left Google to launch a startup incubator and venture capital operation in China, claims that China already has a significant lead on the west in the AI race, in part because privacy is a nonissue. Access to data makes all the difference, he says, and the kind of data being collected in China is qualitatively different from that being collected in the United States. In his words:

Silicon valley juggernauts are amassing data from your activity on their platforms, but that data concentrates heavily in your online behavior, such as searches made, photos uploaded, YouTube videos watched, and posts liked. Chinese companies are instead gathering data from the real world, the what, when, and where of physical purchases, meals, makeovers, and transportation. Deep learning can only optimize what it can see by way of data, and China's physically grounded technology ecosystem gives these algorithms many more eyes into our daily lives.[19]

On a recent trip to Xi'an, I witnessed this transformation first hand. Li Liuyang, a student at Northwestern Polytechnical University, was showing me around the city, a beautiful and deeply historic walled city that was the capital of China for centuries. But the most amazing part of it for me was Li's use of WeChat's Wallet, a service provided by Tencent, one of China's largest Internet companies. Li had no need for cash. He would pick a bicycle, ride it to his destination, buy bottled water from a street vendor, and pay for everything using the app.

I also found that almost no place in China accepts credit cards. Since I don't like carrying cash, I tried to set up the WeChat app for myself, and found that, as a foreigner without a verified identity in China, I was blocked from setting up the Wallet. Tencent, and probably the Chinese government, wants to make sure they know who the real person is behind the online identity. A prepaid credit card can be used anonymously, but not a WeChat Wallet. For the time being, cash still works for most things in China, but this is probably transitory. The machines will eventually track every action of every individual, and to stay anonymous, you will have to drop out of society.

AI SAFETY

A naive answer to these threats is that we should endow AIs with built-in safety envelopes, constraints on their behavior that keep their goals aligned with those of us humans. Indeed, AI safety has become quite a movement, with several institutes and centers established to research the problem. However, the goals of humans are not always so laudable, so aligning the goals of AIs with those of humans may not be such a good strategy. I personally find humans wielding AIs far scarier than the AIs themselves. Nevertheless, as with any powerful technology, finding technological ways to reduce the risk of irreparable harm is essential.

I do not believe that the right approach is to endow the AIs themselves with ethical principles. According to the Dartmouth philosopher James Moor, for a machine to become a "full ethical agent," it must have consciousness, intentionality, and free will.[20] I have argued that while we cannot rule out consciousness and free will for machines, we are nowhere near achieving them. Perhaps more importantly, if they acquire these characteristics, we will never know it for sure. But what we can be sure about is that AIs in the hands of unethical humans will be capable of enormous damage.

Machines are quite good at goal-directed action when the goals are clear and can be formally specified. Feedback control systems are a classic example, as we saw in chapter 5. In many circumstances, however, the goals are not so clear. Norbert Wiener, who worked on feedback control technology for the automatic aiming and firing of anti-aircraft guns, wrote,

> If we use, to achieve our purposes, a mechanical agency with whose operation we cannot interfere effectively...we had better be quite sure that the purpose put into the machine is the purpose which we really desire.[21]

Feedback control endows a machine with "mechanical agency," giving it a measure of autonomy to adjust its own actions to achieve its goal. Imagine the damage that an autonomous anti-aircraft gun can do if its goals are not clear. Even a machine with no AI can be dangerous.

REWARDS AND COSTS

Modern AI algorithms are feedback systems. The AI compares its actions against its goals and adjusts its behavior to get closer. For this to work, goals must somehow be reduced to numbers so that the distance from the goals can be measured. Alternatively, goals can be encoded by a cost function, which is just like a reward function except that the cost is *lowest* when the goals are reached. The naive approach to AI safety, therefore, is to come up with a numerical reward or cost function that encodes what is good for humans.

Stuart Armstrong, a research fellow at the Future of Humanity Institute at Oxford University, in his book, *Smarter Than Us: The Rise of Machine Intelligence*, points out the absurdity of this goal:

> So the task is to spell out, precisely, fully, and exhaustively, what qualifies as a good and meaningful existence for a human, and what means an AI can—and, more importantly, can't—use to bring that about... [and] then code that all up without bugs. And do it all before dangerous AIs are developed.[22]

In their classic AI book, Stuart Russell and Peter Norvig describe what seems like a reasonable reward function for a vacuum robot, to pick up as much dirt as possible.[23] They later pointed out that this reward function can be maximized by a robot that collects dirt and dumps it in order to pick it up again.

One of the challenges in designing autonomous systems, therefore, is finding suitable reward or cost functions. A clever approach called inverse reinforcement learning (IRL) was developed by Russell and his colleagues at Berkeley around the turn of the millennium.[24] In IRL, a machine observes a human expert performing some task, assumes that the human is acting to maximize some unknown reward function, and, based on the actions of the human, builds a reward function.

An auto-pilot system for an aircraft based on IRL, for example, could simply observe the altimeter and the controls while a human pilots the plane. If the altitude drops below the pilot's optimum, the pilot will adjust the controls a little bit. By observing the amount and direction of each control adjustment for each change in altitude, the auto-pilot system can learn a reward function. This reward function reveals that the pilot is trying to keep the altitude within a narrow range. Once it has learned the reward function, the AI could take over and keep the aircraft at the desired altitude. Some researchers have called this strategy "apprenticeship learning."

In 2016, Russell, together with three other Berkeley AI researchers, Dylan Hadfield-Menell, Anca Dragan, and Pieter Abbeel, pointed out two serious problems with IRL.[25] The first is that we don't really want machines to adopt human goals as their own. A human may exhibit a desire for coffee in the morning, but do we want a robot to learn to desire coffee in the morning? Probably not.

The second problem that Russell and his colleagues identified is more subtle. A robot passively observing a human may be an extremely inefficient way to learn. A more interactive approach, one that involves *teaching* rather than just learning, may be much more effective. Consider an auto-pilot system that observes an expert pilot on a long flight. The pilot will keep the airplane very steady, depriving the robot of learning how to appropriately respond to large changes in altitude. This could easily result in a control system that is far from safe.

Russell and his colleagues introduced an improvement that directly speaks to the idea of intelligence augmentation versus AI. They called their technique cooperative inverse reinforcement learning (CIRL). Specifically, instead of optimizing the reward function that the robot learns from a human, the robot learns to optimize the reward reaped by the human. In their formulation, the robot and human cooperate on a task, and the reward reaped by the two is the same. This incentivizes the human to teach the robot. The objective is that the robot should desire coffee for the human, rather than for itself.

An auto-pilot example using CIRL may actually work much more like a human flight instructor. The human pilot, for example, could regulate how much control it delegates to the auto-pilot. Suppose that this amount of delegation could vary from 0 (no delegation, the human pilot has full control) to 1 (full delegation, the auto-pilot has full control). The reward is highest when the human willingly grants 1, fully automatic control, to the machine. It is important that the goal not just be to achieve fully automatic control, because that could be accomplished by sidelining the human, or, worse, killing the human. The goal of the auto-pilot system is to get the human to *willingly* grant it control and continue to grant it control. The auto-pilot system, therefore, will adjust what it does to get the human to give it more control and to continue to grant it control. Any action it takes that causes the human pilot to wrest back control will be negative reinforcement, and any action it takes that causes the human pilot to give it more control will be positive reinforcement. This much more closely resembles the interaction between a human flight instructor and a human student.

That CIRL can be more effective than IRL reinforces a repeating theme in this book. Interaction is more powerful than observation. Both systems can benefit when information flows both ways rather than just one way. But the problem remains that the human's reward function may be nefarious, in which case the resulting AI will be far from safe.[26]

INTERSUBJECTIVE REALITY

The reality is that much of what we humans deem to be real is not. In *Homo Deus*, Harari talks about the "intersubjective reality" created by shared myths such as money and religion. This reality depends on communication, and Harari emphasizes the role that writing has served to strengthen that reality. But Harari's intersubjective reality still requires human brains to put the words to paper and to interpret and act on the words that are written. As we integrate computers into that reality, brains are no longer required.

Consider the stock market, surely a prime example of the power of shared fictions. Harari points out that corporations are a shared fiction. There is no physical-world entity called Google, for example, and yet we can buy stock, another shared fiction, using money, a third shared fiction. It is all a shared fiction, and yet people commit their life savings, another fiction, to it. But today, much of the trading on the stock market is done autonomously by computers. Humans are too slow, and computers can make more money by besting each other by a few milliseconds. As this intersubjective fiction becomes more independent of human brains, what changes? Can we say that the computers themselves have enlisted in this shared fiction?

Is it possible that the computers, trading among themselves, will invent their own intersubjective reality, like Facebook's negotiating bots (see chapter 9)? The 1987 stock market crash (see chapter 10) was widely attributed to a vicious feedback loop fueled by computers realizing an algorithm call portfolio insurance. The algorithm, in effect, codified a belief system, and when the belief system became shared by enough players in the game, disaster! In the case of portfolio insurance, the belief system was designed by humans. But what sorts of belief systems might emerge from AIs? These "beliefs" may not even be meaningfully describable in any human language.

Stock market disasters, of course, do not require computers. Computers were not implicated in the 1929 crash and the ensuing run on banks. In fact, that disaster was also fueled by a shift in the shared belief system, where people started to doubt the soundness of the banks and the stock market. Since banks and stock markets are fictions, the doubt was not only justified, but self-fulfilling, even tautological. The new shared belief system, without banks, simply had fewer and less effective mechanisms for humans to cooperate with one another at large scale. The loss of wealth, another fiction, was inevitable.

Our intersubjective reality may be more fragile than we realize, and disruptions are probably inevitable as more machines interact with each other, bypassing humans. But coevolutionary forces will treat these disruptions as diseases, and some of the AIs will die because they will have killed their symbiotic host. Cambridge Analytica, the British political consulting firm implicated with manipulation of the 2016 US election, is bankrupt. The organizations and people responsible for pathologies may be materially hurt by the diseases introduced by their machine symbionts. Our collective "immune system" will do everything it can to kill pathological machines. Portfolio insurance died in 1987, although mutations of the algorithm have survived. We need to keep an eye on them.

THE BIGGER DANGER

I think we can safely assume that the physical parameters of the human brain will remain relatively static over the next few decades, but the human brain augmented with cognitive machine prostheses will not. Not everyone will have access to the same cognitive prostheses, however. It is possible for a technology to emerge that is so powerful that it could split humanity into haves and have-nots that are so separated that they do not interbreed and cannot understand one another. We are already seeing signs of this with technology-fueled extreme polarization. If this persists for a long enough time, it will inevitably lead to a genetic divide, forking the human race into multiple species. Imagine a world in which a subset of humanity carries AI-enhanced brain implants that change the basic structure of thought enough to make communication with those lacking the implants no more effective than the communication with your dog. This is not a pretty image.

Barrat, Bostrom, Tegmark, and many others warn of the coming superintelligence, the AIs that will eclipse us, leaving behind pathetic, bored, and unemployed shells of humans. I believe the far bigger danger is the humans themselves. Humans will change with the technology, and while we have no evidence of evil in machines, we have plenty of evidence of evil in humans. Humans who use AI as IA, intelligence augmentation, may be the ones to eclipse the rest of us.

We have seen the warning shots. WannaCry and the Dyn attack were human creations exploiting an evolved technological ecosystem for power and financial gain. It may be only a matter of time before we experience the first cyber September 11. We will see AIs become weapons wielded at the hand of nefarious humans. We should fear the humans more than the AIs.

Even well-meaning humans wielding powerful technologies can do an enormous amount of damage. More Americans, and vastly more non-Americans, died as a result of the US reaction to September 11 than died in the event itself. With machine pathologies, there is always the risk that our countermeasures will be worse than the disease, leading to a realization of George Orwell's Big Brother or worse. Big Brother seems to have already happened in China, and the rest of the world may be close behind. In the meantime, we are going to continue to experience stresses with information-fueled partisan warfare, wealth concentration, and abuses of power.

But to me, even scarier is the prospect of humans using AIs to turn our most cherished values into weapons against us. This is what happened with the Russian meddling in the 2016 US election and with the Brexit vote, where filter bubbles stoked the embers of discontent with the goal of creating an entirely irrational result.

What should we do? You may not believe this from what I am saying, but I am an optimist. I believe that the forces governing our development are, to a large degree, beyond our control. But they are themselves a feedback system with its own regulatory mechanisms. If we understand its dynamics, we can nudge it in the right direction. This is the topic of the next and final chapter.

14

COEVOLUTION

===

CHICKENS AND EGGS

Richard Dawkins famously said that a chicken is an egg's way of making another egg. Is a human a computer's way of making another computer? The machines of this book, if we view them as living beings, are creatures defined by software, not DNA, and made of silicon and metal, not organic molecules. Some are simple, with a genetic code of a few thousand bits, and some are extremely complex. Most live short lives, sometimes less than a second, while others live for months or years. Some even have prospects for immortality, prospects better than any organic being. And they are evolving very, very fast. It is not just technology that is changing. We humans are also changing very fast compared to anything found in nature. The way our society works, the way we think, the way we communicate, and, increasingly, even our biology are all in flux.

Are we humans in control of this evolution? Are we truly the masters of the machines? A naive view is what we might call "digital creationism." In this view, we humans use our intelligence to engineer machines in a top-down fashion, like God. A more realistic view is that we are the sources of mutation in a Darwinian coevolution. The mutations we introduce are not entirely random, but in a modern view of evolution, neither are the biological mutations introduced by nature. Much like what happens in nature, most of the technological mutants we bring forth go quickly extinct, while a few grow to occupy a niche, at least for a while, in a continually changing ecosystem.

We have come to depend deeply on technology, to the point that we would not exist without technology. More precisely, we would perhaps exist, but in far fewer numbers and in a form that most of us would find alien and incomprehensible. The "we" of today is a mashup of technology, biology, and culture.

Of the three elements of this mashup, it is technology that is changing fastest. Rapid change can cause problems. Just as disruptions of our gut biome can make us sick, so can technology disruptions. But just as interventions like probiotics can make us healthier, so can technology.

It is natural to fear change. Is AI really an existential threat to humanity? Are we destined to be annihilated by a superintelligent new life form on the planet? Are we destined to fuse with technology to become cyborgs with brain implants that define a new form of quasi-human intelligence? Will we lose control of our machines? If technology is coevolving with humans, then we never really had control. The best we can do is prod the process toward a mutually beneficial symbiosis and deal with the unexpected problems as they arise. Even if we are successful, our future symbiotic selves may not resemble humans of today as much as those who fear change might like.

There are risks, and they are not small. Rapid coevolution is inherently unpredictable, and pathologies are likely to emerge. But we should treat these as pathologies, not as a war with invading aliens. The biggest threat to humanity may not be that the machines will make us irrelevant, but rather that the machines will change the very essence of our being, what it means to be human.

Change is scary, but I, for one, am not nostalgic for any human epoch earlier than now. My lifetime has coincided with the most prosperous and relatively peaceful era in all of human history, and no small part of the credit for that goes to technology. We are not living in Eden, of course, but in many dimensions, the human condition has improved. This does not mean that everything will *continue* to improve, but the better we understand the dynamics of our coevolution with technology, the more likely that it will.

Thinkers such as Vinge, Kurzweil, Bostrom, and Tegmark have written about a runaway feedback loop, where the machines design their own successors, breaking free of any symbiosis with humans. I think they may have overestimated what digital computation can do, but even if they haven't, a more likely outcome is much more powerful (and potentially far scarier) human-machine partnerships. Digital computation is the most potent invention humans have ever come up with, and humans have a horrific track record of using our inventions to perpetrate atrocities on one another. We humans are the scarier part of this partnership.

A FOURTH AGE

Kevin Laland, the evolutionary biologist who appeared in chapters 3 and 13, in his 2017 book, *Darwin's Unfinished Symphony*, identifies three distinct ages in the evolution of humankind, genetic, genetic-cultural, and cultural evolution. Perhaps we have entered a fourth distinct age, one that we might call the "synthetic age."

Laland's first age, genetic evolution, is shared with all other living residents of our planet. It is dominated by biology and by the happenstances of the environment. This phase was, until recently, thought to be dominated by a form of neo-Darwinian evolution where random mutation provides diversity, and environmental and competitive pressures weed out those less able to survive and procreate. As we will see, the story appears to be more complicated in ways that make the evolution of machines look more like that of early biological life.

A key feature of Laland's genetic evolution age is that the creatures evolving are buffeted by environmental events that are entirely out of their own control. A dramatic example, first suggested in 1980 by the father-and-son team of scientists Luis and Walter Alvarez, is the Cretaceous-Paleogene extinction event, where an asteroid or comet strike is believed to have wiped out the dinosaurs and many other species approximately sixty-six million years ago.

The second age, genetic-cultural coevolution, Laland estimates, began some four million years ago and accelerated quite dramatically over the last forty thousand years or so. In this age, humanoids and then humans began to have a strong enough effect on their own living environment that a feedback pattern emerged, where the environment affected the genes (as before), and the genes affected the environment (new to this age). Laland argues that this feedback was enabled by the emergence of culture, which he defines as "the extensive accumulation of shared, learned knowledge, and iterative improvements in technology over time."[1]

In this second age, the switch from a hunter-gatherer society to an agrarian society enabled population growth and demanded social organization. This accelerated the evolution, according to Laland:

Once population size reached a critical threshold, such that small bands of hunter-gatherers were more likely to come into contact with each other and exchange goods and knowledge, then cultural information was less likely to be lost, and knowledge and skills could start to accumulate.[2]

The key feature of Laland's second age is the effect that humans have had on their own environment, becoming "ecosystem engineers." The Dutch evolutionary biologist Menno Schilthuizen, in his book *Darwin Comes to Town*, points out that humans are not nature's first ecosystem engineers. Earlier examples include ants and beavers, who also altered the ecology in ways that then affected their own development. Such feedback loops are fairly common in nature.

Laland describes the third age as follows:

Now we live in the third age, where cultural evolution dominates. Cultural practices provide humanity with adaptive challenges, but these are then solved through further cultural activity, before biological evolution gets moving. Our culture hasn't stopped biological evolution—that would be impossible—but it has left it trailing in its wake.[3]

Why is cultural evolution so much faster than biological evolution? It must be because humans are able to be more intelligent in bringing about mutations. As Turing himself said in 1950,

The survival of the fittest is a slow method for measuring advantages. The experimenter, by the exercise of intelligence, should be able to speed it up.[4]

But Turing was not referring to cultural evolution. He was already referring to what I am calling the fourth "synthetic" age, characterized by what is effectively the emergence of a new life form, one based on silicon rather than carbon. This qualifies as a fourth age because, unlike cultural evolution, where the intelligence is applied to evolve itself, in the synthetic age, human intelligence is being applied to evolve symbionts, the machines, and human intelligence, in turn, is evolving as the machines increase their capabilities. The machine symbionts may later become able to harness their own intelligence to evolve themselves without interaction with humans, as predicted by Bostrom and the others, but this has not really happened yet. If and when it does, we will have entered a fifth age.

All oversimplifications of history that divide it cleanly into distinct phases are flawed, of course. The boundaries between phases are far from clear. But Laland's ages help us distinguish the mechanisms that drive change. The mechanisms driving biological change are clearly not the same as those driving cultural evolution. We could ask whether we should even be using the same word, "evolution," for both. How closely related to Darwin's original idea are these mechanisms? How good is the analogy when we apply the word "evolution" to the development of machines? As it turns out, even in biology, the meaning of the word "evolution" is evolving. But the stalwart constant that sticks with us since Darwin's time is the principle of natural selection, which applies across all these ages.

EVOLUTION ISN'T SO SIMPLE AFTER ALL

Darwin's theory of evolution has, like most scientific theories, evolved with time. Darwin developed his theory long before DNA was understood, so the mechanisms of inheritance and mutation were mysterious. Darwin took for granted, for example, that characteristics an organism acquired during its life could be inherited by its offspring, a view known as Lamarckian inheritance, somewhat unfairly named after the French biologist Jean-Baptiste Lamarck.

The writer David Quammen, in his book, *The Tangled Tree: A Radical New History of Life*, engagingly tells the story of how the theory of evolution has itself evolved. He describes Lamarck as "France's great early evolutionist" who became a bit of a laughing stock for his view of inheritance of acquired characteristics. In Quammen's words,

The most familiar example of such inherited adjustments, which Lamarck himself offered, is the giraffe. The protogiraffe on the dry plains of Africa stretches to reach high foliage, its neck lengthens (supposedly) from the effort, its front legs lengthen too, and therefore (again supposedly) its offspring are born with longer necks and front legs. Lamarckism, in that cartoonish form, has been easy to despise but harder to kill off entirely.[5]

As it turns out, this idea has been "harder to kill off entirely" because it is partly true, although probably not for the giraffe example. Several mechanisms contribute to inherited characteristics, including the passing from generation to generation of the genome of symbiotic microbes (the hologenome), adaptations of the immune system, and epigenetics, particularly proteins bundled with chromosomes that affect gene expression. All of these reflect acquired characteristics and add information to what is passed from generation to generation. In chapter 8, I pointed out that DNA carries nowhere near enough information to create a human, so other mechanisms must exist.

The hologenome has an obvious analogy with the coevolution of humans and machines. When we put an iPad in the hands of our two-year-old kids, they "inherit" ways of interacting with their environment such as swiping the screen and pinch-to-zoom, that are encoded in the "genome" ("codome"?) of our symbiotic machines. These mechanisms, and many more, have integrated with our brains and shape our thinking much more than we realize. If these are "mutations" in humans, they are not encoded in our genome and they have certainly not come about from the classic neo-Darwinian mechanism of random mutation followed by natural selection. Random mutation certainly plays a role, but it is not even close to the whole story.

BEYOND RANDOM MUTATION

Some aspects of relatively new evolutionary theories resemble what we see happening with machines more closely than random mutation. One of the radical new discoveries that Quammen documents is called horizontal gene transfer (HGT). In his words:

The tree of life is more tangled. Genes don't move just vertically. They can also pass laterally across species boundaries, across wider gaps, even between different kingdoms of life, and some have come sideways into our own lineage—the primate lineage—from unsuspected, nonprimate sources. It's the genetic

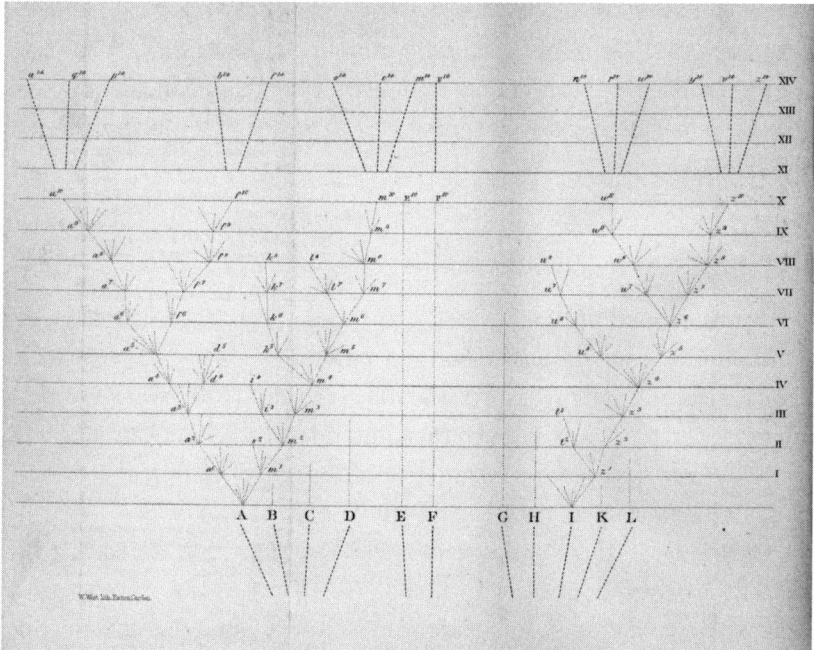

14.1 The only illustration in Darwin's 1859 *On the Origin of Species* was this depiction of the tree of life. The A through L at the bottom are hypothetical unnamed species within some hypothetical genus. The lines on the vertical axis labeled I-XIV each represent a thousand generations. The branching shows variation leading to both extinction and new species.

equivalent of a blood transfusion or (different metaphor, preferred by some scientists) an infection that transforms identity. "Infective heredity."[6]

In Darwin's "tree of life" (see figure 14.1), species fork into subspecies through relatively slow accumulation of small random mutations followed by "survival of the fittest," where "fittest" means most likely to procreate. We have already seen in chapter 9 that this tree is not so simple in that branches can remerge through hybridization. But it turns out to be even more complicated than that.

HGT is apparently common in bacteria and leads to much faster evolution than random mutations. It is now understood to be the primary mechanism for the spread of antibiotic resistance in bacteria. Many biologists assume that HGT played a major role in the early development of life, but there is also evidence that it played a role later in much more advanced life forms, including humans. According to Quammen, researchers have identified about one percent of the human genome that very likely got itself inserted through HGT mechanisms in the last few million years.

BACTERIAL SEX

At least three mechanisms for HGT have so far been identified. They are called *transformation*, *conjugation*, and *transduction*. The first to be discovered, transformation, dates back at least to the 1920s, when a British physician, Fred Griffith, noticed that a harmless bacterium could change suddenly into a virulent form that would cause pneumonia, a leading cause of death in those days.

Much later, in the 1940s, the biologist Oswald Avery, working at the Rockefeller Institute in New York, identified DNA as the material from which genes and chromosomes are made, and he found that free-floating DNA from dead bacteria could lead to the kinds of transformations that Griffith had observed. Live bacteria absorb the dead genetic material through their cell membrane and edit it into their own DNA. The transformation mechanism that Avery identified, which was later called "infective heredity," turned out to be fairly common in bacteria. Keep in mind that this was nearly ten years before the landmark publication in 1953 of Watson and Crick's paper describing the double-helix structure of DNA. Avery was repeatedly nominated for the Nobel Prize, which he never received.

In 1946, a twenty-one-year-old researcher, Joshua Lederberg, took a leave of absence from medical school at Columbia to work at Yale University under the guidance of microbiologist Edward Tatum. At Yale, in less than two years, he met and married another student of Tatum's, Esther Miriam Zimmer, identified a second HGT mechanism that he called conjugation, coauthored with Tatum a paper published in *Nature* on conjugation, wrote and filed a PhD thesis, and accepted an assistant professorship in genetics at the University of Wisconsin at Madison. That was a busy two years.

In the mechanism that had been identified by Avery, transformation, a bacterium takes up genetic material left behind by other bacteria that have died. At Yale, Lederberg showed that gene transfer could occur between living bacteria as well. He was not quite twenty-two when his paper with Tatum appeared in *Nature* showing that temporary cell fusion and exchange of genetic material must be occurring. They dubbed the process conjugation and called it a "sexual process."

In 1951, working with his graduate student Norton Zinder in Madison, Lederberg identified a third HGT mechanism that they called transduction, where viruses carry DNA from one strain of bacteria into another. He and his wife, Esther Lederberg, who received her PhD from the University of Wisconsin in 1950, later collaborated to identify a specialized version of transduction that is less random. In 1958, at age thirty-three, Joshua Lederberg shared the Nobel Prize with Edward Tatum and George Beadle for his work on genetics. In that year, he moved to Stanford, where he founded the department of genetics.

Lederberg later made significant contributions to computer science. In the 1960s, he played a central role in the development at Stanford of an influential AI program called Dendral, which helped organic chemists to identify unknown organic molecules. This was a good old-fashioned AI (GOFAI)–style expert system based on encodings of knowledge of chemistry in the form of production rules (see chapter 4).

HORIZONTAL CODE TRANSFER

HGT upended evolutionary biology. The mechanisms identified by Avery and Lederberg result in faster evolution than the neo-Darwinian mechanism of random mutation followed by natural selection. What biologists mean by "random mutation" here is mutation that is caused by extraneous factors, not by any process that is part of the normal biological processes of the organism. They can occur, for example, due to x-rays or environmental toxins. Although we have seen in chapter 11 that the notions of randomness and causation are far from simple, some sources of mutations, such as x-rays, are clearly extraneous. For some time, many biologists believed that such extraneous mutations were the dominant source of variation in evolution.

The vast majority of such random mutations, however, affect cells that are not germline cells, those involved in procreation, such as eggs and sperm, and hence do not get passed on to offspring. Moreover, the vast majority of random mutations are deleterious and therefore will also not be passed along. If this were the only source of variation, then evolution would likely be much slower than it is.

HGT provides faster random mutations, but it also creates opportunities for less random, more targeted mutations, where strains with a beneficial gene can transfer those genes to entirely different strains of bacteria. This discovery shook the foundations, calling into question the concept of a species and the tree of life, at least with respect to bacteria.

Later discoveries show that HGT occurs throughout nature, not just in bacteria. Human DNA contains significant segments that appear to come from bacteria and even viruses. Carl Zimmer reports in the *New York Times* that scientists have found some 100,000 elements in the human DNA that probably come from viruses.[7]

It is tempting to draw an analogy to computer viruses, such as those that are picked up by opening documents sent in a phishing email message. A notable example is the Melissa virus, which infected Microsoft Word. You receive an email from a friend with a message like "Here is that document you asked for; don't show it to anybody else." You open the file, and a macro embedded in the file accesses your contacts in Microsoft Outlook and sends those people email messages similar to the one you just received. Is this analogous to transformation, where Microsoft Word is absorbing "genetic" material (in the form of macros) from its environment and splicing it into its own "genome"? This is not a very good analogy, however. The infected Microsoft Word does not pass along the mutation to any offspring, but rather passes it along to peer individuals. The mechanism is more analogous to the spreading of a disease like the common cold.

A much better analogy can be found in the software development process itself. Software engineering is the discipline of creating new strains of digital machines. The code that is engineered is the "DNA" of the machines. But a software engineer rarely starts from scratch. It is much more common to start with a working program and modify it. A software engineer, therefore, is the source of mutation in a neo-Darwinian "tree of code."

But like the tree of life, the tree of code gets tangled. A software engineer will frequently splice into the code of one program fragments from another. The engineer is acting like the viruses in transduction that carry DNA from one cell to another. We can call this process "horizontal code transfer."

Analogous to transformation, the first HGT mechanism, an engineer will pick up code fragments that are not living, in that they are not part of working programs but are found on the Internet or in libraries of software components. The engineer will splice those fragments into a new program. Beneficial components, ones that have proven useful in many programs, are more likely to get spliced into the "codome" of a new program.

Engineering, like evolution, is about creating artifacts and processes that have never before existed. We engineers tend to think of our role in this process as that of creator, an intelligent designer who, in a top-down fashion, coerces matter and energy to do our bidding. We take great pride in the outcome, our creation, our invention, like our children. But also like our children, we have less control over the outcome than we imagine. Kevin Kelly, whom we met in chapter 2, cites what he calls the "adhocracy" of Wikipedia as evidence that we don't need much top-down design to get fantastic outcomes.[8] We only need a little. We are arguably more like mediators of mutation in an evolutionary process than top-down intelligent designers.

Moreover, our own thinking, during the process of engineering software, evolves along with the machines we build. The software supporting software development shapes the process, splicing memetic material into our cognitive "genome" ("memome"?). The software tools we use to create software change our minds, which in turn changes the software we write.

TOP-DOWN INTELLIGENT DESIGN?

The philosopher Daniel Dennett, in his 2017 book, *From Bacteria to Bach and Back*, makes the case that the human mind, our consciousness, languages, and cultures, are the result of an evolutionary process. He is not talking about the brain and its biological structure and processes, but rather is saying that the mind emerges from more than biology. Dennett is defending and elaborating on the earlier controversial position famously put forth by Richard Dawkins in his 1976 book, *The Selfish*

Gene, where Dawkins coined the term "memes" for cultural artifacts and ideas, drawing an analogy between their propagation in human culture and Darwinian evolution. From Dawkins:

I think that a new kind of replicator has recently emerged on this very planet. It is staring us in the face. It is still in its infancy, still drifting clumsily in its primeval soup, but already it is achieving evolutionary change at a rate which leaves the old gene panting far behind.…The new soup is the soup of human culture. We need a name for the new replicator, a noun which conveys the idea of a unit of cultural transmission, or a unit of imitation.[9]

That noun is "meme."

Dawkins had quite a few detractors who did not like his analogy with biology, but Dennett argues that even some of the most fervent detractors espoused, using other words, essentially the same theory that ideas, culture, and languages propagate via a neo-Darwinian natural selection, where, in Dennett's words, "fitness means procreative prowess." The post–neo-Darwinian mechanism of HGT may be an even better analogy because mutation of ideas is not, mostly, randomly caused by factors entirely extraneous to culture.

Drawing an analogy between biological evolution and evolution of machines is easier even than between biological evolution and evolution of memes because digital machines are more like biological living beings than memes are (see chapter 2). Memes do not have an autonomous existence in the physical world, independent of human brains, but machines do.

Dennett, however, falls short of identifying today's technology as part of his evolving ecosystem. On the contrary, he points to digital technology and software as a canonical example of an opposite kind of design from evolution, what he calls "top-down intelligent design." I will shorten this to "TDID" to avoid repeating the phrase too often. Dennett argues that TDID is less effective than evolution at producing complex behaviors, contrary to a religious position held by some that the complexity of life proves the existence of God.

Dennett points to an elevator controller, observing that every contingency, every reaction, every behavior of the system is imposed on it by the cognitive engineer who designed it. This is partly true, but many aspects of the design have been heavily influenced by prior technology developments, so even a modest elevator controller is the result of an evolutionary process. Moreover as machines go, an elevator controller is a rather simple one. For more complex digital and computational behaviors, like those in Wikipedia, a banking system, or a smart phone, it is hard to identify any cognitive being that performed anything resembling TDID. These systems evolved through the combination of many components, themselves similarly evolved, with engineers introducing both mutations and horizontal code transfer. Together with decades-long iterative design revisions with many failures along the way, we get a Darwinian process of mutation and natural selection.

Dennett argues that, unlike biological beings, the parts of a digital design have no yearnings for resources, nothing driving them forward, no purposes or reasons, and that they are just reactive automata. But this isn't a useful distinction because many alternative designs and mechanisms died along the way, and the ones that survived did so for Darwinian reasons, because they were able to propagate. Barring unsound teleology, propagation is also the closest that biological evolution gets to having a purpose. The propagation of machines is facilitated by the very concrete benefits they afford to the humans that use them, for example by providing those humans with income and hence with food and the ability to procreate.

Viewing software as "top-down intelligent design" falls victim to the same tendency that Dennett criticizes, the homunculus in the brain, a little man or committee that observes and drives the decision making of the human mind. In contrast, a coevolutionary stance says that software evolves in much the same way that bacteria evolve, through a goal-less coevolution with humans driven by its own Darwinian reward functions, survival and propagation. The tendency to see these designs as TDID is anthropocentric, a tendency that we, as humans, find naturally difficult to avoid. We do not like seeing our mental cognitive processes themselves as cogs in a relentless purposeless evolution. But is this what they are?

FACILITATORS OR INVENTORS?

Dennett even applies his TDID principle to artifacts that are far too complex to have been designed this way:

To take the obvious recent example of such a phenomenon, the Internet is a very complex and costly artifact, intelligently designed and built for a most practical or vital purpose: today's Internet is the direct descendant of the Arpanet, funded by ARPA (now DARPA, the Defense Advanced Research Projects Agency), created by the Pentagon in 1958 in response to the Russians beating the United States into space with its Sputnik satellite, and its purpose was to facilitate the R&D of military technology.[10]

This is an oversimplification of the Internet. ARPA funded the development of a few of the protocols that underlie the Internet, but even these protocols emerged from many failed experiments at methods for getting computers to interact with one another.[11] Moreover, ARPA and DARPA had little to do with most of what we recognize as the Internet today, including web pages, search engines, YouTube, and so on. As I pointed out in the previous chapter, Tim Berners-Lee, creator of the web, laments what it has become. Much of the Internet evolved from the highly competitive entrepreneurial dog-eat-dog ecosystem of Silicon Valley and the collaborative minds of thousands of contributors to the standards that make it as robust as it is today.

The computer scientist and entrepreneur Danny Hillis, referring to the Internet, writes,

Although we created it, we did not exactly design it. It evolved. Our relationship to it is similar to our relationship to our biological ecosystem. We are codependent and not entirely in control.[12]

Further digging himself in, Dennett demurs,

All of this computer R&D has been top-down intelligent design, of course, with extensive analysis of the problem spaces, the acoustics, optics, and other relevant aspects of the physics involved, and guided by explicit applications of cost-benefit analysis, but it still has uncovered many of the same paths to good design blindly located by bottom-up Darwinian design over longer periods of time.[13]

Dennett does not see that "computer R&D" is actually more like Dawkin's memes than like TDID. Humans are more facilitators than inventors, and as Dennett notes about culture,

Some of the marvels of culture can be attributed to the genius of their inventors, but much less than is commonly imagined…[14]

The same is true of technology.

EVOLUTIONARY MULTIPLIERS

Although Dennett overstates the *amount* of TDID in technology, there can be no doubt that human cognitive decision making strongly influences its evolution. At the hand of a human with a keyboard, software emerges that defines how a new machine strain reacts to stimulus around it, and if those reactions are not beneficial to humans, the strain very likely dies out. But this design is constructed in a context that has evolved. It uses a human-designed programming language that has survived a Darwinian evolution and encodes a way of thinking. It puts together pieces of software created and modified over years by others and codified in libraries of software components. The human is partly doing design and partly doing random mutation, horizontal code transfer, and simple husbandry, "facilitating sex between software beings by recombining and mutating programs into new ones."[15] So it seems that what we have is a *facilitated* evolution, facilitated by elements of TDID and conscious deliberate husbandry. There are many examples of facilitated evolution in nature, including, for example, the Cambrian explosion (see chapter 2); human husbandry of farm animals, domestic pets, and crops;[16] evolution of animals and plants to adapt to human urbanization;[17] and the development of antibiotic-resistant bacteria via HGT.

As with any evolutionary process, competition for resources plays a role in the evolution of machines, and death and extinction are natural parts of the process. The success of Silicon Valley depends on failure of startup companies as much as it depends on their success. Software competes for a limited resource, the attention and nurturing of humans that is required for the software to survive and propagate. Consider the browser wars of the 1990s, where many attempts at programs for viewing content on the Internet succumbed to competition in acts of deliberate and systematic killing. Having been caught by surprise by the emergence of the web, starting around 1995, Microsoft built Internet Explorer into all Windows systems, free of charge, in a deliberate attempt to kill off the competing browsers. Today, few browser species survive.

Wikipedia and Google are spectacular multipliers of our human cognitive abilities, but they are not themselves TDIDs. Although their evolution has most certainly been facilitated by various small acts of TDID, they far exceed as affordances anything that any human could have possibly designed. They have coevolved with their human symbionts.

Dennett observes that collaborating humans vastly surpass the capabilities of any individual human. Humans collaborating *with technology* further multiplies this effect. Technology itself now occupies a niche in our (cultural) evolutionary ecosystem. It is still relatively primitive compared to humans, much like our gut bacteria, which facilitate digestion. Technology facilitates thinking.

PARASITIC OR SYMBIOTIC?

Dennett takes on AI and most particularly deep learning systems, calling them "parasitic." He focuses on their mechanics, noting that while they can classify images, for example, those images have no meaning to them. They "parasitically" derive any meaning from humans. In his words,

Deep learning (so far) discriminates but doesn't notice. That is, the flood of data that a system takes in does not have relevance for the system except as more "food" to "digest."[18]

This limitation evaporates when these systems are viewed as symbiotic rather than parasitic. In Dennett's own words, "deep learning machines are dependent on human understanding."

Dennett notices a similar partnership between memes and the neurons in the brain:

There is not just coevolution between memes and genes; there is codependence between our minds' top-down reasoning abilities and the bottom-up uncomprehending talents of our animal brains.[19]

For the neurons in our brain, the flood of data they experience also has no "relevance for the system except as more 'food' to 'digest.'" An AI that requires a human to give semantics to its outputs[20] is performing a function much like the neurons in our brain, which also, by themselves, individually have nothing like comprehension. It is an IA, intelligence augmentation, not an AI.

DUMBING DOWN

Today, there is a lot of hand wringing and angst about AI. Dennett raises one common question:

How concerned should we be that we are dumbing ourselves down by our growing reliance on intelligent machines?[21]

Are we dumbing ourselves down? It doesn't look that way to me. This does not mean we are out of danger. Far from it. Again, from Dennett:

The real danger, I think, is not that machines more intelligent than we are will usurp our role as captains of our destinies, but that we will overestimate the comprehension of our latest thinking tools, prematurely ceding authority to them far beyond their competence.[22]

I believe there are far bigger dangers than this one. First, IA in the hands of nefarious humans and governments is a scary prospect indeed. Second is that the machines will change our thinking, as they already have, through the creation of filter bubbles and echo chambers.

Evolutionary pressures may tend to accentuate the fragmentation of information through a phenomenon that evolutionary biologists call the Baldwin effect, named after the American philosopher James Mark Baldwin (1861–1934). Under this effect, an organism's ability to learn new behaviors during its lifetime affects its reproductive success and will therefore have an effect on the genetic makeup of its species through natural selection. Today, a search engine that acquires enough "knowledge" of me to tune its results to what I want to hear is more likely to survive and propagate in the ecosystem of search engines, which compete for advertising dollars. As the search engine learns, its reproductive prowess improves, thereby reinforcing the development of machines that fragment human thinking. As it learns, it creates for me an ever smaller echo chamber, feeding me only the information I want to see. And its progeny will even more effectively isolate our progeny from each other.

A third danger bigger than the one Dennett cites is that the machines will shed their dependence on humans and that we will lose control. This is the danger that Bostrom, Tegmark, and others focus on. It is true that there have been moderately successful experiments where programs learn to write programs, and it seems inevitable that the machines will continue to get better at designing themselves. This fear is real, but it may be that we never really were in control, so losing control is not the essential issue. If the machines are evolving in a Darwinian way, then the best we can do is nudge the process. We cannot really control it, but through policy and regulation, we may be able to slow or even prevent undesirable outcomes.

Dennett's final words are optimistic:

If our future follows the trajectory of our past—something that is partly in our control—our artificial intelligences will continue to be dependent on us even as we become more warily dependent on them.[23]

I share this optimism, but also recognize that rapid coevolution, which is most certainly happening, is extremely dangerous to individuals. Rapid evolution necessarily involves a great deal of death. Both technologies and memes will fall by the wayside as the symbiogenesis evolves. *Coevo-lution* means that both parties, the humans and the technologies, will change. Even if this remains symbiotic, the results can be dramatic. The resulting humans may be very different from the humans of today.

ENDOSYMBIOSIS

Analogies can be useful intuition pumps, to use the words of Dennett, but they are risky. I am drawing an analogy between evolution of digital technology and both biological evolution and Dawkins's memetic evolution. Just as with memes, for digital technology, mutation and natural selection both occur, where humans provide the mechanisms for both. It is not just an analogy. The parallel with life *is* an analogy, but that parallel is not so important. What is important is that we understand the mechanisms of change, and not oversimplify by vilifying individual technologists each time we discover a pathology in the evolving ecosystem. If these mechanisms of change truly were TDID, then vilifying the engineers may be justified. But the mechanisms are more complex. We are all complicit, for example, in the shape of the ecosystem that determines whether a technology strain either succeeds and propagates or fails and goes extinct. The engineers act like the viruses in HGT, transporting "genetic" material from one technological strain to another. But the rest of society overuses antibiotics, thereby creating an ecosystem that naturally leads to antibiotic-resistant bacteria.

Although we *are* facing the possibility of the machines affecting our genes, today, the human side of the coevolution is still mostly memetic rather than biological. Our ability to mutate technology, to engineer new strains, evolves like Dawkins's memes, considerably pushed along by the technology itself, which provides the software and hardware that we routinely use to engineer new software and hardware. A strong feedback loop forms, where technology causes memetic mutation, and the memes cause technology mutation.

It turns out, however, that there is an even stronger and scarier analogy with biology. The mutual dependence we have with technology is a symbiosis, and symbiosis can lead to an even stronger source of mutation than HGT. Biologists call this source of mutation *symbiogenesis*. It is where an entirely new and more complex life form emerges from a fusing of the partners in a symbiosis. Symbiogenesis is also called *endosymbiotic theory*, and an endosymbiosis is a symbiosis where no partner can live without the other—like its weaker cousin, an obligate symbiosis—but the partners have fused to become one, where one lives within the tissues of the other.

The biologists David Smith and Angela Douglas give cows as an example of an endosymbiosis. Cows, they say, are "forty-gallon fermentation tanks on four legs."[24] Lynn Margulis, who deserves much of the credit for our current understanding of symbiogenesis, describes cows this way:

> Cows ingest grass, but they never digest it because they are incapable of cellulose breakdown. Digestion in cows is by microbial symbionts in the rumen. The rumen is a special stomach, really an overgrown esophagus, that has changed over evolutionary time. Cows that lack rumens don't exist; cows (and bulls) deprived of their microbial symbionts are dead.[25]

A cow is not a creature that contains microbial symbionts. Rather, the symbionts are no less part of the cow than the rumen itself. Without the symbionts, there is no cow.

Human dependency on technology has not quite reached this stage, in the sense that humans would continue to exist without technology, albeit in far fewer numbers. But the strength of the codependence keeps increasing, and it is not farfetched that we will reach a point where what we mean by "a human" includes the technologies without which that human cannot live.

EVOLUTIONARY DISCONTINUITY

The relationship between a cow and its gut microbes is asymmetric. The microbes are physically much smaller and biologically simpler than the cow. The relationship between humans and technology today is also asymmetric. Digital artifacts are far simpler than our brains, and we at least have the illusion of being in control, using technology as a tool. This asymmetry will likely decrease over time as technology gets more sophisticated, and the resulting symbiosis could become first an obligate symbiosis and eventually an endosymbiosis.

In biology, there are less asymmetric endosymbioses than that of a cow. The human cells in our bodies, as well as those in all plants and animals, very likely emerged as an endosymbiosis of simpler creatures. These cells are quite different from those of bacteria, which lack mitochondria, chloroplasts, and a nucleus. Those organelles have their own enclosing membranes, and most biologists today believe that they evolved from independent creatures that fused to form today's cells. Biologists call cells with such organelles *eukaryotes* and distinguish them from *prokaryotes*, which, like bacterial cells, have no such internal structure. Eukaryotes evolved from a symbiosis between prokaryotes. The importance of this step cannot be overstated:

The largest evolutionary discontinuity on this planet is not between animals and plants; it is between prokaryotes (bacteria without membrane-bounded nuclei) and eukaryotes (all the others made of cells with membrane-bounded nuclei). The detailed story of this huge discontinuity is connected to the origins of species.[26]

The evolutionary biologist Ernst Mayr called the emergence of eukaryotes "perhaps the most important and dramatic event in the history of life."[27] The merging of humans with technology, if it happens, will be equally momentous.

Neo-Darwinian evolution of humans may have slowed because we produce fewer offspring than we used to, so there are fewer mutations per parent, and those offspring are more likely to survive and reproduce. Better health care, clean water, and safe food attenuate the effect of natural selection. Put differently, the memetic evolution that keeps us from drinking the water pooled in the gutter affects the gene pool, illustrating the Baldwin effect.

Further evolution of the human genome may, in the future, occur more through genetic engineering than through random mutation or HGT. George Dyson speculates:

Are we using digital computers to sequence, store, and better replicate our own genetic code, thereby optimizing human beings, or are digital computers optimizing our genetic code—and our way of thinking—so that we can better assist in replicating them?[28]

But even without genetic engineering, humanity may change through a symbiogenesis with technology. Our pacemakers and insulin pumps on the biological side, and our banking, transportation, and communication systems on the cultural side, may be the precursors of symbionts without which some future form of humans will become less able to procreate. It is not hard to imagine, for example, a world in which sex never leads to pregnancy and humans lose the ability to become pregnant that way.

Endosymbiotic theory is relatively young. Lynn Margulis was twenty-nine years old when in 1967 she published "On the Origin of Mitosing Cells" under the name Lynn Sagan (she had married and then divorced the famous science popularizer Carl Sagan). Her title is a clear bow to Darwin's 1859 *On the Origin of Species*. In this paper, she resurrected what many biologists considered to be a wacky idea first advanced by the Russian botanist Konstantin Mereschkowski, who, in the early 1900s, suggested that eukaryotic cells evolved from a symbiosis between distinct prokaryotic cells. It was Margulis who put the theory on a sound biochemical footing. In a highly influential book written later with her son, Dorion Sagan, she writes,

We believe random mutation is wildly overemphasized as a source of hereditary variation. … Rather the important transmitted variation that leads to evolutionary novelty comes from the acquisition of genomes. Entire sets of genes, indeed whole organisms each with its own genome, are acquired and incorporated by others. The most common route of genome acquisition, furthermore, is by the process known as symbiogenesis.[29]

The genome of a mitochondria within a human cell is distinctly differ-ent from that in the nucleus of the cell. Both sets of genes are inherited, although the mitochondrial genes only from the mother. Mereschkowski and Margulis's hypothesis is that, far in the past, one cell ingested another, and instead of digesting it, hijacked its functions to make it part of a new type of cell.

Some human lives already depend on technologies incorporated into our bodies, for example pacemakers. But a pacemaker is not inherited by offspring. If we reach the point where human newborns are routinely augmented with technological prostheses, or where procreation is always mediated by machines, we will have entered a new era for biological life. More dramatically, is it possible that we humans will become the mito-chondria of the technium, organelles that perform a vital function in the larger being but that cannot live on their own? Today, this is the stuff of science fiction.

But other scary scenarios loom closer. An endosymbiosis forms with the fusing of two life forms into one. Technology today cannot live with-out us humans, so although our dependence on it is not absolute, its dependence on us is. Will we become like the gut bacteria of technium, able to live outside the host, but only at the cost of very poor health? Gut bacteria do not fare well on their own. Or worse, will we become the para-sites or pathologies of the technium, doomed to be subjugated or even annihilated? We are already seeing "machine medicine" and "machine immune systems" improving, where software self-repairs and AIs expunge malware. What if we humans become tantamount to malware?

WILL WE BE ECLIPSED?

Our dependence on technology has been steadily growing, a trend that seems likely to continue, but technology would go extinct overnight without the help of humans. Is it likely to shed that dependence on us? For this to happen, the machines will need to operate, procreate, and evolve without the help of humans.

In 2018, a team of researchers at the University of Toulouse and the University of York created a program that could write programs to play old Atari video games credibly.[30] Their program generated random mutations and then simulated natural selection. Their technique was itself evolved (via horizontal code transfer) from earlier work that evolved programs to develop certain image processing functions.[31] In principle, these projects and many other fledgling efforts on automatic coding show that if the machines somehow figure out how to keep themselves running without the help of humans, they could evolve their software without the help of humans. Moreover, their evolution would be using a method, natural selection, that is known to be effective at producing very sophisticated beings.

The Atari game–playing programs that emerge from the Toulouse-York evolutionary process, however, are far less effective than programs based on deep learning. The Toulouse-York team admits this, saying that the main advantage of their technique is that the resulting programs are more explainable (see chapter 6). The game-playing strategies can be read (by humans) from the evolved programs. Such an advantage, however, is irrelevant if there are no humans demanding explanations.

Evolution is a form of learning. To the extent that there is a distinction between evolution and learning, evolution governs what emerges at birth and learning governs what emerges during life. In biological systems, both forms of acquired capability are passed on to offspring, the first primarily through genetics, and the second primarily through memetics.

In both cases, information that passes from one generation to the next over a noisy channel, according to the Shannon channel capacity theorem (see chapter 8), carries only a finite number of bits. In *machine* learning, versus human learning, a finite number of bits is all there is, at least today, and hence capabilities that a technology acquires during its "life" can be passed on *perfectly* to its offspring. Lamarckian inheritance is a reality for digital technology. For biological creatures, the story is less clear, however, because some information is carried from generation to generation by the "thing in itself," the continuous biological process that is some four billion years old. This information is not limited to a finite number of bits.

Moreover, by the Baldwin effect, the introduction of machine learning into a wider variety of technological artifacts will enhance their procreative prowess. Their ability to learn during life will make them more adaptable to changing environmental conditions, which makes them more likely to survive and propagate. For example, if humans were to decide someday to kill off some strains of technology, only those that can adapt to this hostile environment will survive and propagate. We are already seeing human-created regulations and laws prohibiting certain kinds of technologies, and we are seeing adaptation in technology strains to survive these laws. Some technologies also succumb to pathologies, becoming extinct because their weak security makes them too vulnerable to viruses and worms. The inability to adapt can doom a species. For example, in December 2018, Google announced that they would kill Google+, citing new vulnerabilities to malware that were not worth the cost to fix. Google+ was, apparently, insufficiently adaptive.

IMMORTALITY

Today, the state of an executing computer program can be copied, stored, and restored perfectly with astonishingly high confidence. This is possible because the essential properties of the program are digital. Inessential properties, such as the temperature of the chips running the program, cannot be perfectly copied, but those properties do not define the being. Digital traits can also be perfectly passed on to offspring.

However, many digital technologies are not *completely* digital. Robots, for example, are not robots unless they have a physical presence able to interact with the physical world. Self-driving cars are not self-driving cars unless they have wheels and can move through physical space. The robotics researchers Paul Fitzpatrick, Giorgio Metta, and Lorenzo Natale, in a paper entitled, "Towards Long-Lived Robot Genes," lament,

Robot projects are often evolutionary dead ends, with the software and hardware they produce disappearing without a trace afterwards.[32]

As machines become more embodied (see chapter 7), their inheritance mechanisms will inevitably become less perfect.

In chapter 8, I pointed out that any being that is completely defined by a digital code can, in principle, become immortal. Nature, however, has given us no immortal beings. In fact, evolution does not even *favor* longevity, much less immortality! Peter Godfrey-Smith, who appeared in chapter 2 with his study of octopuses, has pointed out that every evolutionary advantage comes with a cost, and evolution favors advantages that help early in life, before and during procreation, at the expense of costs incurred later in life, after procreation. This explains why evolution has not and probably never will deliver immortality. As we age, we pay for the strong body we once had. Embodied machines will likely similarly never develop immortality. They too will age and die.

INTELLECTUAL SIDELINING

Even if embodied robots fail to eclipse humans, to the extent that intelligence can be accomplished in a purely digital way, humans still may be intellectually sidelined. In a 2016 TED talk, Sam Harris, whom we met in chapter 10, made the case that intelligence is information processing, and that the information-processing abilities of our machines will continue to improve. He concludes that it is only a matter of time before they eclipse us. Harris is not alone in drawing such a conclusion. Nick Bostrom, Max Tegmark, and Kevin Kelly have all written similar predictions.

Despite my argument in chapter 1 that intelligence does not lie on a linear scale, the prediction is hard to refute. It is certainly possible for the machines to continue to improve in *all* relevant dimensions of intelligence, in which case they could certainly sideline us. However, these writers do not make any distinction between digital information and nondigital information. If the latter is essential, then we have not yet invented the technology that will eclipse us.

My argument in chapter 8, that cognition (probably) is not digital and algorithmic, can, perhaps, just slow down our progress toward doom. As we learn more about the neuroscience of intelligence, it will become easier to make machines that *do* include the right sorts of processes to match and exceed any cognitive function in humans. Our brains are irrefutable proof that it is possible to make intelligent machines, since nature has done so. Is it farfetched to assume that only biochemical machines driven by human DNA are capable of such intelligence? Could the concept of embodied cognition save us? Digital machines will never have human bodies.

In chapters 11 and 12, we saw that interaction is more powerful than computation. In chapter 7, we saw that interaction with the physical world is central to cognition. Although, today, machines are far less embodied than humans, they are interacting with the physical world more every day. Kai-Fu Lee, whom we met in chapter 13, points out that China's Internet and AI infrastructure already penetrates deeply into the physical world. Such eyes and ears are the first step toward an embodied cognition.

The second step is for the machines to manipulate the physical world. Just as Facebook's machines can experiment with user interface designs (see chapter 11), learning from the reactions of the users, Tencent's machines can experiment with physical actions. How does placement of bicycle stands affect mobility in a city? How does pricing of services affect where people go? How can users be incentivized to leave scooters where they are most likely to be picked up and used again? As these computer systems close the feedback loop, affecting the physical world and measuring its reaction, will this reafference (see chapter 5) inevitably result in self-awareness and human-like intelligence? My guess, and it is just a guess, is that the intelligence that will emerge will not resemble human intelligence much at all. But this is far from reassuring.

There is another weakness in the argument that humans will be sidelined, although this weakness is also far from reassuring. The doomsday scenarios compare humans of *today* to machines of *tomorrow*. But humans will change too, and indeed we already are changing. Our cognitive and physical beings are already intertwined with the machines, and this codependence and integration is only going to accelerate. This does not necessarily result in a less scary picture, however.

SOULLESS MACHINES?

Despite the emergence of AI-generated art (see chapter 10), perhaps we can derive solace from a soulful sensation that only humans can possibly create and appreciate poetry, music, and dance. Douglas Hofstadter expresses this sensation this way:

Many educated people believe that although a machine may now or someday be able to do a creditable job of acting like a person, any machine's performance will always remain lackluster and dull, and that after a while this dullness will always show through. You will simply have no doubt that the machine is unoriginal, that its ideas and thoughts are all being drawn from some storehouse of formulas and clichés, that ultimately there is nothing alive and dynamic—no *élan vital*—behind its façade.[33]

The utterances of Tay on Twitter, however, were anything but dull (see chapter 10).

Reducing art to neuroscience, Steven Pinker says,

The real medium of artists, whatever their genre, is human mental representations. Oil paint, moving limbs, and printed words cannot penetrate the brain directly. They trigger a cascade of neural events that begin with the sense organs and culminate in thoughts, emotions, and memories.[34]

If the essence of art is "human mental representations," then by definition, machines cannot participate. They are not human. The purpose of art becomes the conveyance of these mental representations from one human to another.

We have a word for words that are especially economical and effective at conveying mental representations. We call these words "poetry." But even a poem is imperfect. The thoughts it triggers in your mind will not match those in the mind of the poet no matter how poetic the words are. Often, the power of poetry lies in its ambiguity and its ability to adapt to the individual, to trigger powerful and personal emotional thoughts in a human whose cognitive world is very different from that of the poet.

We have already seen that technology gives an artist a richer palette and more versatile media. It has never been the case that the art is created by a paintbrush, but a good paintbrush can make a big difference. And the most effective paintbrush is the one designed to work well with the human hand and human eye. As machines get ever more deeply synergistically intertwined with our human world, they will inevitably provide us with more media for creativity. Their role in the human soul, therefore, is not to replace it with dry objectivity. Instead, they have real potential to enrich our artistic lives by providing us with entirely new kinds of paintbrushes. Recall from chapter 10 that Plato, in *Timaeus*, asserts that if we understand the mechanisms that cause a human action, that action becomes soulless, one for which we cannot hold the human accountable. When we assume that the actions of a machine will be soulless, it is perhaps because we assume that the mechanisms behind those actions are explainable. But as we saw in chapter 6, modern AI programs yield behaviors we cannot explain. This may be the reason that artists have taken note and are starting to use AI as an art medium.

But we can go even further. Today's software is digital and algorithmic. The physical world, on the other hand, is (probably) neither digital nor algorithmic, and it can exhibit both nondeterminism and chaos, both of which make behaviors fundamentally unpredictable. Future machines may harness both of these to produce genuine delight in their human symbionts.

ETHICAL TECHNOLOGY

Digital technology today is a tsunami swamping human culture. It is changing our political systems, economies, and social relationships. It is redefining our intellectual lives, changing how we pursue science, anthropology, art, and literature. It creates fabulous wealth and opportunity while devastating entire careers. It informs and misleads, unites and divides, and empowers and paralyzes. It unleashes free speech and enables ubiquitous surveillance. And that is just today. What about tomorrow?

There are enormous opportunities and risks. How can we mitigate the risks and maximize the opportunities? Many educators believe that the answer is to teach ethics in engineering and computer science schools. If this is indeed a solution, then a corollary is that bad outcomes are the result of unethical actions by one or more individuals. But given the complexity of socio-technical interactions and the coevolution thesis of this book, this corollary is probably invalid. It is analogous to the assumption that if each neuron in the brain is operating normally, then mental illness cannot emerge. Under that assumption, mental illness could be treated by identifying the rogue neuron or neurons and killing them. I don't think any credible psychiatrist or neuroscientist is pursuing such a route.

While it is certainly important that engineers behave ethically, teaching ethics is not a panacea. Even if we could get every technology developer to behave ethically, an unrealistic goal, pathologies will still emerge. Many of the detrimental effects that we have seen are unintended and unanticipated consequences of well-meaning actions. If we overemphasize ethics, we could end up just vilifying scapegoats without really improving anything.

Today, the only effective principle guiding technology development seems to be the pursuit of profit. This is a strong motivator, and it stimulates creativity, but it is a blunt instrument, and history has shown that it must be regulated. To do better, we have to first understand the complex dynamics of an evolving socio-technical culture.

Some people use the term "digital humanism" for a human-centric study of technology. It is imperative for intellectuals of all disciplines to step up and take seriously this intellectual challenge. Our limited efforts to rein in the detrimental effects of technology have been, so far, mostly ineffective, underscoring our weak understanding of the problem. The privacy laws in the United States and Europe, for example, are not accomplishing their objectives. And it is not even clear that the privacy goals can be met even if all human participants behave ethically, an unrealistic expectation.

WHAT SHOULD WE TEACH THE YOUNG?

Humans have a handicap compared to machines. Because of the digital nature of their knowledge, everything that a digital machine learns can be copied nearly instantaneously to another similar machine. Humans, on the other hand, start from scratch and have to go through a decades-long painful and imperfect process of knowledge transfer that we call "education." This handicap, however, is also an opportunity. If we start early, focusing young minds on the hard questions of digital humanism, perhaps we have a chance. After all, it is the next generation that will both drive innovation and bear the brunt of the mistakes.

Traditionally, a well-educated person is one with knowledge of language, history, and science, and the skills to manipulate formal systems like mathematics and computer programs. Today, it seems that wisdom takes a back seat to skills and knowledge of facts. This has proved valuable, making our young more employable. Skills and facts, however, are increasingly becoming better handled by machines, so perhaps these are not the best choices for what to emphasize in the future.

It is clear that our young should study technology, but, I believe, not primarily to enhance their job prospects, but rather to enhance their understanding of the society they are growing up in. It is a nice side benefit that, in the short term, it *will* enhance their job prospects, but the durable value comes from developing a deeper understanding of the tectonic forces that will make all those skills obsolete and their knowledge superfluous.

Instead of just teaching kids how to write programs in Python, for example, we should also introduce them to Guido van Rossum at CWI in The Netherlands, the creator of the original Python, and to the open-source community that has grown up around Python. They should develop an understanding of the sociology of Python and open-source software.

We should introduce our young to ideas around privacy, using their own tools—Snapchat, WeChat, Instagram, and Facebook—as illustrations. Privacy is a fascinating philosophical conundrum and a relatively recent concept. Studying the technology around privacy can lend insights into what it really means for humans. We should help them understand the dynamics of viral spread of ideas. This is a very different teaching agenda than teaching them how to write programs that sort numbers, the focus of most introductions to computing today.

Sadly, most educators do not do the sort of teaching I have in mind. For most of my career as a professor, neither did I. It would not have occurred to me that Python has a history, that it emerged from the mind of a single creative individual and then evolved into an entire ecosystem of technological species, most of which will go extinct. To me, Python was a Platonic fact about the world and may as well have always existed. To me, all programming concepts had this character.

I have a very different view today, but I spent many years spreading a profound misunderstanding of technology, one that locks our young into a hopeless acceptance of technological "facts" about the world. We think we are empowering them by giving them the skills to get jobs, but we are actually shutting them in to today's facts and making them vulnerable to a changing world. It is perhaps ironic that the technologists of tomorrow need to be our strongest humanists.

PUBLIC POLICY

Humans need heroes. We like to single out brilliant individuals and give them Nobel Prizes and credit them as inventors and entrepreneurs. Every time we do this, we ignore thousands of other individuals, each of whom was essential to the outcome. On the flip side of the coin, when technology leads to bad outcomes, we like to single out the villains. Hence, we drag Silicon Valley executives in front of Congress and threaten to break up their companies. Assigning blame to greedy capitalists may make us feel good, but it has little effect on future technology outcomes. Attacking capitalism itself will affect future societal outcomes, even if not technology outcomes, but most of us probably will not like those outcomes. The twentieth century tried that experiment. So what should we do to prevent bad technology outcomes?

Under digital creationism, the purpose of regulation is to constrain the individuals who develop technology. Under coevolution, the purpose of regulation is to nudge the process of technology development. Under digital creationism, bad outcomes are the result of unethical actions by individuals, for example by blindly following the profit motive with no concern for societal effects. Under coevolution, bad outcomes are the result of procreative prowess. Technologies that succeed are those that more effectively propagate. The individuals we credit with creating those technologies certainly play a role, but so do the users of the technologies. Should we establish policies to constrain those users?

Consider privacy laws. I believe these have been ineffective because they are based on digital creationism as a principle. These laws erroneously assume that changing the behavior of corporations will be sufficient to achieve privacy goals. A coevolutionary perspective understands that users of technology will choose to give up privacy even if they are explicitly told that their information will be abused. We are repeatedly told exactly that in the fine print of all those privacy policies we don't read.

I don't have a concrete proposal that will effectively improve personal privacy. I am not even sure what it means to improve personal privacy. I value freedom, so individuals should be free to give up their own personal privacy. Most of the people that I talk to tell me that they have nothing to hide so they don't mind giving up their privacy. But what if the collective actions of many such individuals lead to an Orwellian state, as it has in China?

I believe that, as a society, we can do better. I'm not sure how to prevent an Orwellian state (or perhaps, worse, a corporate Big Brother). But I am sure that we will not do better until we abandon digital creationism as a principle.

NOTES

CHAPTER 1

1. In a 2014 review article, Alcock, Maley, and Aktipis say, "Evolutionary conflict between host and microbes in the gut leads microbes to divergent interests over host eating behavior. Gut microbes may manipulate host eating behavior in ways that promote their fitness at the expense of host fitness."

2. A study by geneticist Carl Bruder of the University of Alabama at Birmingham and his colleagues has determined that there are sometimes variations in the genes of identical twins, specifically in the number of copies of a particular gene (Casselman, N.D.). These differences can result in different phenotype, but differences manifest even when the DNA is identical.

3. Dawkins, *Blind Watchmaker*, p. 5.

4. Dawkins, *Blind Watchmaker*, pp. 185–186.

5. Patricia Churchland explains nicely how foresight provides selection advantages in many animals, not just humans (Churchland, 2013).

6. Rogers and Ehrlich, "Natural Selection and Cultural Rates."

7. Handwerk, "Gut Bacteria May Be Controlling."

8. Sigmund, *Exact Thinking*, pp. 146–147.

9. Pinker, *Enlightenment Now: The Case for Reason, Science, Humanism, and Progress*, p. 296.

CHAPTER 2

1. Dyson, *Turing's Cathedral*, pp. 308, 313, 325.

2. Dyson, *Darwin among the Machines*, p. 121.

3. Hobbes, *Leviathan*, p. 7.

4. Langton, *Artificial Life*, p. 1.

5. A good summary of field of artificial life is given by Aguilar et al. (2014).

6. Langton, *New Definition*.

7. Aguilar et al., "Past, Present, and Future."

8. von Neumann, "General and Logical Theory."

9. Emmeche, *Garden in the Machine*, p. x.

10. Lee, *Plato and the Nerd*.

11. Maturana et al., *Autopoiesis and Cognition*, p. xvii.

12. Dennett, *Intuition Pumps*, p. 4.

13. Oddly, small amounts of amino acids were found in samples brought back from the moon by the Apollo missions. According to NASA, scientists don't think the organic matter came from life on the moon. Instead, they identify several possible sources. The samples may have been contaminated by either the missions to the moon or in the handling of the samples back on earth. Alternatively, rocket exhaust from the lunar modules contains precursor molecules that could have turned into amino acids during analysis in the lab. Similar precursor molecules are found in the solar wind—a thin stream of electrically conducting gas continuously ejected from the surface of the Sun. These precursors could again have turned into amino acids during manipulation of the samples in the lab. Finally, amino acids have been found in asteroid fragments that occasionally fall to Earth as meteorites. The lunar surface is frequently bombarded by meteorites and could have acquired amino acids from asteroids as well. Of course, these meteorites could explain the origin of amino acids on Earth as well, but that would leave open the question of how the amino acids emerged in the asteroids, so experiments like the Miller-Urey experiment remain interesting.

14. Wolchover, "New Physics Theory."

15. England, "Statistical Physics."

16. Kauffman, *Origins of Order*.

17. Parker, *Blink of an Eye*.

18. Retrieved on May 29, 2018.

19. Dennett, *Intuition Pumps*, p. 98.

20. Dickinson, *Complete Poems*, p. 312.

21. Pinker, *Blank Slate*, p. 423.

22. Pinker, *Blank Slate*, p. 424.

23. Lichtman, "Can the Brain's Structure Reveal?"

24. Lichtman et al., "Big Data Challenges."

25. Mitchell, *Machine Learning*, p. 2.

26. https://blog.wikimedia.org/2018/04/24/new-data-center-singapore/.

CHAPTER 3

1. Bratsberg and Rogeberg, "Flynn Effect and Its Reversal."

2. Laland, *Darwin's Unfinished Symphony*, pp. 29, 209.

3. Laland, *Darwin's Unfinished Symphony*, p. 224.

4. Stringer, "Brain Size Has Increased."

5. McLuhan, *The Gutenberg Galaxy*; McLuhan, *Understanding Media*.

6. Russell, *Human Compatible*.

7. Harari, *Homo Deus*, p. 397.

8. Harari, *Homo Deus*, p. 311.

9. Harari, *Homo Deus*, p. 2.

10. Harari, *Homo Deus*, p. 158.

11. Harari, *Homo Deus*, p. 131.

CHAPTER 4

1. You can view the result at https://youtu.be/ZX564BRcOdo.

2. Nietzsche, *Will to Power*, p. 283, emphasis in the original.

3. Wittgenstein, *Tractatus Logico-Philosophicus*, 5.6.

4. Chesterton, *G. F. Watts*.

5. Carmena et al., "Learning to Control."

6. Haugeland, *Artificial Intelligence*.

7. http://www.masswerk.at/elizabot/.

8. Weizenbaum, "ELIZA."

9. Dreyfus and Dreyfus, "Limits of Calculative Rationality."

10. Kelley, "Optimal Flight Paths."

11. Bryson et al., "Steepest-Ascent Method."

12. Dreyfus, "Artificial Neural Networks."

13. Lee and Messerschmitt, *Digital Communication*; Barry et al., *Digital Communication*.

14. Rumelhart et al., "Learning Representations."

15. Giles, "The GANfather."

CHAPTER 5

1. The idea that motor efference copies play a key role in speech production is argued by Tian and Poeppel (2010), who say "anticipated auditory consequences of planned motor commands" form an essential part of speech production.

2. Black, "Stabilized Feed-back Amplifiers."

3. Godfrey-Smith, *Other Minds*.

4. A nice history is given by Grüsser (1995).

5. Pinker, *Blank Slate.*

6. Wiener, *Cybernetics*, p. 97.

7. Rosenblueth et al., "Behavior, Purpose, and Teleology."

8. To be effective, the delayed speech has to be played into headphones so that it is louder than undelayed speech that propagates from the vocal tract to the ears through the bones and tissue of the head. A classic work on this topic is by Lee (1950). More recently, the UCLA psychologist Donald MacKay teases apart effects of the age of the speaker and the differences between stuttering and distorted sounds (MacKay, 2005).

9. Instantaneous feedback and its relationship to causality is a deep subject with some fascinating implications, beyond the scope of this book, in mathematics, computer science, and philosophy. In mathematics, feedback appears in the form of fixed-point theories, where the interesting properties of a function F are captured by the values x that satisfy $x = F(x)$. These values are called "fixed points." An equation like $x = F(x)$ is circular because the unknown x depends on itself. So how can it become known? Does a fixed point x exist? Is it unique? Systems that are modeled by such equations where there is more than one fixed point can exhibit nondeterministic behavior. In computer science, a family of programming languages known as "synchronous reactive languages" are feedback systems where a program specifies a self-referential relation that needs to be satisfied, and the job of the execution engine is to find the behavior that satisfies this relation (Benveniste and Berry, 1991). Classes of programs where such a behavior exists, is unique, and can be deduced in a finite number of steps are said to be "constructive" (Berry, 1999). On the philosophical side, instantaneous feedback has connections with intuitionistic logic. In intuitionistic logic, "truths" are facts that can be deduced from prior facts in a constructive way. For my take on some of these issues, see Lee (2014) and Lee (2016).

10. Dennett, *Elbow Room*, p. 32.

11. Daniel Dennett has coined the rather awkward word "heterophenomenology" to describe a study of cognition that does not depend on introspection but is based rather on externally observable phenomena (Dennett, 2013). But heterophenomenology fundamentally breaks any feedback loop. Its key property is that the observer is not also the observed. This may make it impossible to develop a deep understanding of anything, if understanding itself intrinsically requires feedback. In chapter 12, I will reveal that prohibiting interaction, relying on observation alone, indeed ties our hands behind our backs.

CHAPTER 6

1. Kosinski et al., "Private Traits and Attributes."

2. Wang and Kosinski, "Deep Neural Networks."

3. Ribeiro et al., "Why Should I Trust You?"

4. Ribeiro et al., "Why Should I Trust You?"

5. Simonite, "Google Photos Remains Blind."

6. Cooper et al., "Predicting Pneumonia Mortality."

7. Caruana et al., "Intelligible Models for HealthCare."

8. Wachter et al., "Right to Explanation in GDPR."

9. Danziger et al., "Extraneous Factors in Judicial Decisions."

10. Kahneman, *Thinking Fast and Slow*.

11. Taleb, *Black Swan*.

12. Taleb, *Black Swan*.

CHAPTER 7

1. Putnam, "Psychological Predicates."

2. Thelen, "Grounded in the World," p. 5.

3. Thelen, "Grounded in the World," p. 7.

4. The word "algorithm" comes from the name of the Persian mathematician, astronomer, and geographer, Muhammad ibn Musa al-Khwarizmi (780–850), who was instrumental in the spread of the Arabic system of numerals that we all use today.

5. Thelen, "Grounded in the World," p. 8.

6. Clark and Chalmers, "Extended Mind."

7. Clark, *Supersizing the Mind*.

8. James Gleick, *Genius: The Life and Science of Richard Feynman*, p. 409.

9. Clark, *Supersizing the Mind*.

10. Sapolsky, *Behave*, p. 588.

11. Sapolsky, *Behave*, p. 588.

12. Carmena et al., "Learning to Control."

13. Clark, *Supersizing the Mind*, pp. 3–4.

14. Brooks, "Artificial Life."

15. Bongard et al., "Resilient Machines."

16. Clark, *Supersizing the Mind*, p. 57.

17. Hofstadter, *Strange Loop*, p. 193.

18. Clark, *Supersizing the Mind*, p. 59.

CHAPTER 8

1. Gribbin, *Alone in the Universe*.

2. Tegmark, *Life 3.0*, location 4038 in the Kindle edition.

3. Parfit, *Reasons and Persons*.

4. Dennett, *Consciousness Explained*, p. 430, emphasis in the original.

5. Hofstadter, *Strange Loop*, p. 257.

6. Hofstadter, *I Am a Strange Loop*, p. 315

7. Shannon, "Mathematical Theory."

8. Chapter 7 of *Plato and the Nerd* (Lee, 2017) has a gentle introduction to the problem of quantifying information. In brief, entropy is a measure of the expected information gained from an observation of something. Determining the entropy requires having a probability measure, which quantifies what we do not know about the something being observed. For an introduction to probability measures, see chapter 11 of that book. If the something being observed has a finite number of possible outcomes, then its entropy tells us how many bits, on average, will be required to represent an outcome. On the other hand, if the something being observed has a continuum of possible outcomes, then no finite number of bits can encode its outcomes. Hence, while its entropy still quantifies its information content, this information measure does not have units of bits. It is still possible to compare the information contained in one something to the information contained in another something, but the information is not representable with a finite number of bits.

9. Wright, "Relative Importance of Heredity."

10. We could argue that a finite number of bits can arbitrarily closely approximate anything that requires an infinite number of bits to encode. However, I can show that an arbitrarily close approximation may completely fail to have essential properties of the thing being approximated. One argument is given in my previous book, section 10.3, where I show an example of a nondeterministic system that can be arbitrarily closely approximated by deterministic ones. Determinism is surely an essential property, and it is a property of Turing machines. Another argument is based on the loss of expressiveness when we restrict our reasoning to countable sets. Mathematically, a world in which space is discrete is actually much more complicated and difficult to model than one that admits continuums. A simple illustration of this is the difficulty defining even simple geometric shapes, such as circles, when you restrict your mathematical universe to countable sets. If space is discrete, then the set of locations in space is countable even if space is infinite. It turns out that Diophantine equations, which describe many shapes, have bizarre and chaotic properties in countable sets. A circle is a simple example of a shape that is the solution to a Diophantine equation. For example, a circle in Euclidean space can be defined as the set of (x, y) solutions to the equation $x^2 + y^2 = 1$, which is a Diophantine equation. As it happens, a number of mathematicians have made entire careers studying the properties of rational solutions to Diophantine equations (Hartnett, 2017). The rational numbers form a countable set, and it turns out that these sets of solutions exhibit some very weird and chaotic properties, very much unlike the real-number solutions to the same equations. Geometry is much simpler in a world of continuums than in a digital world.

11. Clarke, *2001: A Space Odyssey*.

12. Brown, *Origin: A Novel*.

13. Hofstadter, *Strange Loop*, p. 194.

14. That the same process can be modeled as discrete or continuous is supported by the wave-particle duality in quantum mechanics. However, accepted particle models rely on a space-time continuum and therefore are not purely discrete. Models that discretize space and time are not widely accepted and lack experimental support.

15. See, for example, Shapiro (2012).

16. Copeland (2017) has a nice section on common misunderstandings of the Turing-Church thesis.

17. Peter Wegner, a computer science professor at Brown University, has argued that interactive programs can do more than algorithms (Wegner, 1997). At the Workshop on Foundations of Interactive Computation, held in Edinburgh in April 2005, a panel was held and a summary was published (Wegner et al., 2005). Quoting from the summary, "the Church-Turing thesis is commonly interpreted to imply that Turing machines model all computation. It is a myth that the original thesis is equivalent to this interpretation of it (Goldin and Wegner, 2005). In fact, the Church-Turing thesis only refers to the computation of *functions*, and it specifically excludes interactive computation."

18. Chaitin, "Real Numbers"; Chaitin, *Meta Math*.

19. Chaitin, *Meta Math*.

20. These can be listed because the set of all texts in any fixed written language is countable.

21. It is easy to construct an infinite sequence of valid yes-no questions. For example, let the first question be "Is one a whole number?" Let the second question be "Is the answer to the first question 'yes'?" Let the third question be "Is the answer to the second question 'yes'?" And so on.

22. A more rigorous form of the argument would use Cantor's diagonalization technique. The text would describe that diagonalization technique, and thereby describe a number that is not in the list of all describable or nameable numbers.

23. For a gentle introduction to formal languages and the difference between countable and uncountable sets, see chapters 8 and 9 of *Plato and the Nerd* (Lee, 2017).

24. Chaitin, "Real Numbers."

25. Moreover, the arguments for digital physics using the Bekenstein bound and the holographic principle are based on a flawed interpretation of the Bekenstein bound that fails to recognize the distinction between the entropy of a discrete random variable (which represents information in bits) and the entropy of a continuous random variable (which does not represent information in bits). It is a flawed mapping of the physics concept of entropy onto Shannon's information theory. See chapters 7 and 8 of *Plato and the Nerd* (Lee, 2017).

26. Chaitin, "Real Numbers."

27. Dyson, *Turing's Cathedral*.

28. Chaitin, "Real Numbers."

29. Lee, *Plato and the Nerd*, p. 180.

30. I have defended this conjecture more completely in chapter 10 of my previous book, *Plato and the Nerd*. The argument relies on a mathematical result that I first reported in 2016 (Lee, 2016). First, note that a noiseless measurement apparatus must be deterministic. That is, given the same inputs, it must always yield the same outputs. Loosely, my result shows that any apparatus measuring a sufficiently rich combination of discrete and continuous behaviors can always be pushed into nondeterminism, where the outputs can be different given the same inputs. A more precise statement is that any (sufficiently rich) set of deterministic models of the physical world that includes both discrete and continuous behaviors is incomplete. The set does not contain its own limit points. A measurement apparatus that can give different measurements for the same physical object is clearly noisy. Hence, there are only two ways to avoid this noise. The first is to disallow discrete behaviors, to assume that they do not exist in the physical world and hence need not be measured. This is the antithesis of digital physics, so it obviously won't help the dataist cause. The second is to disallow continuous behaviors, or in other words to assume digital physics. Assume that the world is actually discrete, and the hypothesis that the world is actually discrete becomes scientific. This kind of circular reasoning has not traditionally held much sway in science, but perhaps, given the power of feedback and self-reference discussed in chapter 5, circular reasoning will become more respectable in the future. Only then can we rationally accept the digital physics hypothesis.

31. As of 2016, the most precise time measurements ever made had an accuracy on the order of zeptoseconds, a full twenty-three orders of magnitude bigger than the Planck time. One zeptosecond is a trillionth of a billionth of a second. One hundred trillion billion Planck times fit within one zeptosecond.

32. Wheeler, "Unity of Knowledge."

33. Rovelli, *Order of Time*, p. 84, emphasis in the original.

34. Rovelli, *Order of Time*, p. 140.

35. Rovelli, *Order of Time*, p. 84.

36. Rovelli, *Order of Time*, p. 90.

37. A deep analysis of this question is given by Dodig-Crnkovic (2006).

CHAPTER 9

1. Note that computers today really are built on an analog substrate even if digital physics is ultimately true. The design of a transistor does not rely on digital physics. Our best models of the underlying physics do rely on quantum phenomena, so there is an element of discreteness, but the models operate in a time and space continuum. It's all about electrons sloshing around in silicon under the influence of electric fields, a distinctly analog process.

2. Kelly, *Inevitable*.

3. See the YouTube video at https://www.youtube.com/watch?v=E8Ox6H64yu8.

4. The Turing test, proposed in 1950 by Alan Turing, is a way to determine whether a computer program exhibits intelligent behavior equivalent to or indistinguishable from a human(Turing, 1950). In this test, a human evaluator observes natural-language conversations between another human and a computer that is programmed to generate human-like responses. The evaluator would be aware that one of the two partners is a computer but would not know which one. Turing said that if the evaluator cannot reliably tell the computer from the human, then the computer is said to have passed the test.

5. See the CNET story at https://www.cnet.com/how-to/what-is-google-duplex/.

6. Pollock and Samuels, "Jade Helm Exercise."

7. Vincent, "Lyrebird."

8. Baraniuk, "'Creepy Facebook AI' Story."

9. Ford, *Rise of the Robots*.

10. Bostrom, "History of Transhumanist Thought."

11. Bostrom says that Julian Huxley, brother of Aldous Huxley, author of *Brave New World*, first used the word "transhumanism" in *Religion Without Revelation* (Huxley, 1927). However, I was unable to find the word "transhumanism" in that book.

12. Good, "Speculations Concerning Ultraintelligent Machine."

13. Vinge, "Technological Singularity."

14. Goldberg, "Robot-Human Alliance."

15. Ford, *Rise of the Robots*.

16. Legg and Hutter, "Universal Measure of Intelligence."

17. Hart, "Wall of Lava Lamps."

18. Armstrong, *Smarter Than Us*.

19. Armstrong, *Smarter Than Us*.

20. Lucas, "Minds, Machines, and Gödel."

21. See chapter 9 of *Plato and the Nerd* (Lee, 2017) for my take on Gödel's incompleteness theorems. See *Gödel, Escher, and Bach* (Hofstadter, 1979) for a delightful development of the true importance of these theorems.

22. Lucas, "Minds, Machines, and Gödel."

23. Hofstadter, "Can Inspiration Be Mechanized?," pp. 18–34.

24. Chalmers, *Conscious Mind*.

25. Penrose, *Emperor's New Mind*, p. 30.

26. Penrose identifies two classes of noncomputable phenomena in physics, chaos and nondeterminism, and asserts that he cannot see any way that either of these can give rise to consciousness. He cites sources of nondeterminism in classical mechanics (simultaneous multiple collisions), relativity (a phenomenon called "cosmic censorship"), and quantum mechanics. He dismisses the first by assuming it away, choosing to "[ignore] the multiple collision problem" (Penrose, 1989, p. 219). I presume

he felt comfortable doing this under the assumption that the brain has no collisions. But he seems to have missed a related form of nondeterminism in classical mechanics that arises from metastable states, which I will examine in chapter 11. I will argue that this form of nondeterminism *can* play a central role in creativity and free will, which are arguably two key attributes of a conscious being.

Regarding chaos, Penrose is looking specifically at smooth chaotic functions operating in a continuum and argues that for these to be used to create consciousness, one has to assume infinite precision measurements are possible. However, his conclusion is questionable. The weather, like consciousness, is a phenomenon of the physical world. The weather harnesses chaos with no need to make infinite precision measurements. Penrose is confusing the map and the territory here. Measurements are needed for constructing maps, not for realizing the "thing in itself" (to use Kant's term). So chaos may also play a role similar to metastable states, since like nondeterminism, chaos makes the future unpredictable. See *Plato and the Nerd* (Lee, 2017) for a more in-depth discussion of these issues.

27. Lake et al., "Human-Level Concept Learning."

CHAPTER 10

1. See http://obvious-art.com/.

2. Vincent, "Three French Students."

3. In practice, reliability remains a challenge. Sorensen and Reinke (2018) report, for example, that a system that has been deployed to nearly a million vehicles by Nissan to apply automatic braking for collision avoidance sometimes disables itself at unexpected times.

4. Awad et al., "Moral Machine Experiment."

5. Harari, *21 Lessons.*

6. Doyle, *Free Will.*

7. Gaudiano, "One Key Factor"; Leland and Rubinstein, "Evolution of Portfolio Insurance."

8. Git is an open-source version control system originated in 2005 by Linus Torvalds, a Finnish-American software engineer who is famous for another open-source project with a huge impact, the Linux kernel, which is the heart of the Linux, Android, and Chrome operating systems. To use Git, a programmer first "clones" the repository onto her own computer. When the programmer is ready to edit the program, she "pulls" the latest version of the program from the repository, edits it, and then "pushes" the changes back to the repository. If the push is successful, her version becomes a new latest shared version of the program. In a sequence of steps that is probably more complicated than it really needs to be, the push is merged with changes from other programmers and checked for conflicts before being allowed. A conflict can occur, for example, if another programmer has pushed changes to the same part of the program that she has changed. If there is a conflict, she must revise her changes, taking into account the other programmer's changes,

possibly overriding them or further modifying them in some way. It is a chaotically democratic process.

9. Many deep learning programs use common toolkits that could make them easy to recognize because the code will contain signature patterns.

10. In a TED talk in February 2010, Harris lays out his argument that morality has a solid rational basis in logic and science. The basis of his argument is the statement that in comparing two situations that are identical except for one factor, the situation that produces less suffering in conscious beings is more moral than the other. Mathematicians would call this a partial order relation. It is "partial" because there can be two situations that are "incomparable." For example, if one factor leads to less suffering in situation A and the other factor leads to less suffering in situation B, then neither factor is more moral than the other. One problem with this argument is that the most moral situation, one with no suffering, can be achieved by having no conscious beings. Another difficulty is that suffering and consciousness are hard to measure or even to define. Harris argues that suffering and consciousness can be given a basis in science, although we have a great deal more to learn about neuroscience and psychology before we reach that point. Once we construct such a partial order with scientifically grounded notions of consciousness and suffering, I wonder whether the resulting partial order will form what mathematicians call a lattice, where any pair of situations has a unique least upper bound (a unique more moral situation that is no more moral than it needs to be to be more moral than both situations) and a unique lower bound (a unique less moral situation that is no less moral than it needs to be to be less moral than both situations). I seriously doubt that it will be a lattice, in which case, moral disagreements will persist forever even with a solid scientific grounding.

11. Harris, *Free Will*, p. 1.

12. Harris, *Free Will*, p. 4.

13. Harris, *Free Will*, p. 9.

14. Libet et al., "Time of Conscious Intention"; Libet, "Unconscious Cerebral Initiative."

15. Haynes, "Decoding and Predicting Intentions."

16. Fried et al., "Internally Generated Preactivation"; Haggard, "Decision Time."

17. Harris, *Free Will*, pp. 7–8.

18. Recall from chapter 5 that an "efference" is a motor signal from the central nervous system to the peripheral muscles. Hence, a "fake efference" is a similar signal, but one that does not cause any muscles to move. A "reafference" is the signal picked up by your senses, such as your ears, that is the consequence of actions you have taken, such as speaking. You speak, it creates sound, and your ears hear that sound. A "fake reafference" foregoes the actual physical sound. Your "inner voice" is an example of a fake reafference.

19. Note that the phrase "at the time of the Big Bang" is problematic. A naive view of the Big Bang might seem to place the origin of the universe at a point in time,

some fourteen billion years ago, but modern theories of cosmology posit that time and space themselves did not exist before that point. Viewed another way, if we were to travel backward in time toward the Big Bang, we would never reach it because our clocks would keep slowing, as predicted by Einstein's general theory of relativity, so that the Big Bang would become a receding horizon. See Muller (2016). Under such a model, we cannot state that there was a determination between the impossible and the possible outcomes at the time of the Big Bang. The possible outcomes just are, and any imagining of impossible alternatives is just pure fiction. Oddly, that imagining is occurring in the very world where what is being imagined is impossible and has always been impossible, the world of my brain as I write this book.

20. I apologize profusely to both Sam Harris and Daniel Dennett for putting words in their mouths here. I hope I am correctly representing their positions, but in case I am not, I implore the reader to understand that this *my interpretation* of what they have written, and as I've stated clearly earlier in the book, my interpretation is not and cannot be theirs.

21. Dennett, *Elbow Room*.

22. Harris, *Free Will*.

23. A hypothetical kind of machine called a nondeterministic Turing machine provides a useful conceptual framework for a class of machines that may ultimately prove much more capable than today's computers. Conceptually, a nondeterministic Turing machine is one that has more than one possible action at some given state. One way to think of the operation of such a machine is that when it is in a state where more than one action is possible, it executes all possible actions simultaneously, thereby exploring a much bigger solution space quickly. Such machines are still digital and algorithmic, but they can, in principle, solve problems that would take today's computers far too long to solve. They afford a tractable solution to complex problems, but they are not fundamentally able to solve problems that a Turing machine cannot solve, given enough time. Quantum computers, which remain laboratory curiosities as of this writing, are, in principle, capable of such simultaneous exploration of many possible solutions. They are often compared to nondeterministic Turing machines, but whether they are equivalent in some fundamental way remains controversial.

24. The "law of the excluded middle" is an axiom of classical logic that asserts that any sentence must be either true or false. A system is "nondeterministic," logically, if it is "not deterministic." By the law of the excluded middle, a statement that a system is deterministic must be either true or false, and if it is false, then the system is nondeterministic. There is, however, a form of logic called "intuitionistic logic," that rejects the law of the excluded middle. In intuitionistic logic, a sentence is true only if there is a proof that it is true, and it is false only if there is a proof that it is false. Under this logic, a statement that a system is deterministic may be neither true nor false. Intuitionistic logic replaces the classical law with a constructive principle, which states that truth or falsehood are consequences of constructive demonstrations of that truth or falsehood. Only under intuitionistic logic does Harris's position admit the possibility of a mechanism possessing free will.

25. Harari, *Homo Deus*, p. 282.

26. Kastrenakes, "Microsoft Made a Chatbot."

27. Vincent, "Twitter Taught Microsoft's Chatbot."

28. For example, in Brscic et al. (2015), a group of researchers in Japan document a study of the behavior of children toward social robots placed in a shopping mall. The children would sometimes speak offensively, block their way, and even kick or punch the robot. In an earlier study, Bartneck et al. (2005) repeated the famous Milgram experiment with robots rather than human subjects and determined that people were far more willing to administer damaging shocks to robots than to other humans. They coined the term "robot abuse" for this phenomenon.

29. Harris, *Free Will*, p. 28.

30. Fremont et al., "Control Improvisation."

31. Donzé et al., "Machine Improvisation."

32. Akkaya et al., "Control Improvisation."

33. Hofstadter, *Strange Loop*, p. 20.

34. United States v. Grayson, 1978, and Morissette v. United States, 1952.

35. Under US law, a corporation, as distinct from its associated human beings (owners and employees), has at least some of the legal rights and responsibilities of humans. In particular, corporations can be held accountable for things for which the owners and employees of the corporation are not accountable.

36. Vigna and Casey, *Age of Cryptocurrency*, p. 222.

37. Sapolsky, *Behave*.

CHAPTER 11

1. Davies, "Sleeping Tesla Driver."

2. Russell, "Notion of Cause."

3. Hitchcock, "What Russell Got Right," p. 53.

4. Norton, "Causation as Folk Science," p. 34.

5. A nice collection of essays on the deep influences of the notion of causality on language is found in *Causation in Grammatical Structures* (Copley and Martin, 2014).

6. https://www.3ammagazine.com/3am/the-causal-revolutionary, retrieved October 15, 2018.

7. Pearl and Mackenzie, *Book of Why*, p. 349.

8. Pearl and Mackenzie, *Book of Why*, p. 89.

9. Pearl and Mackenzie, *Book of Why*, p. 79.

10. Pearl and Mackenzie, *Book of Why*, p. 84.

11. Pearl and Mackenzie, *Book of Why*, p. 269.

12. See Norton (2007). Another example with similar properties is given in Dhar (1993).

13. Norton, "Causation as Folk Science," p. 26.

14. Norton goes a step further asking us to consider a slightly different scenario, where the ball starts on the outskirts of the hill and we push it with just enough force that it reaches the top of the hill and stops. If we push too hard, it will go over the top of the hill. If we don't push hard enough, it will not reach the top of the hill and will fall back down. But if we push it with the Goldilocks force, just right, it will stop at the top, stay there for an arbitrary amount of time, and then spontaneously roll down the hill again sometime in the future.

For this scenario to work, Norton points out that the shape of the hill is important. If the hill is a perfect hemisphere, then with the Goldilocks force, it will take infinite time for the ball to reach the top of the hill. It will keep slowing down as it approaches the top, but it will never actually reach the top. But there are many other hill shapes where the ball reaches the top in finite time. Norton develops one particular example where the hill drops by a distance $h = (2/3g)r^{3/2}$, where r is the horizontal distance from the center of the hill and g is the force of gravity. This choice makes the math work out particularly simply.

Newton's laws are usually assumed to be time reversible. They work the same way whether time moves forward or backward. So pushing the ball up the hill to come to a stop can be understood as a time reversal of the previous scenario. But, as Norton points out, there are subtleties. If we push the ball up the hill and it stops at the top at some time T_1, then at any time $T_2 \geq T_1$, it could begin rolling down the hill again. We end up with many possible time-symmetric behaviors, most of which are not time reversals of each other.

15. For example, Marino (1981), Kinniment (2007), and Mendler et al. (2012).

16. Ditlevsen and Samson, "Introduction to Stochastic Models."

17. Earman, *Primer on Determinism*, p. 21.

18. Lee, *Plato and the Nerd*.

19. An "input" is a modeling concept that does not really exist in the physical world. It is a stimulus to a component on which the component has no direct influence. An input is imposed by the environment, and in violation of Newton's third law, we neglect the reaction that the component has on the source of that input. If Newton's third law is correct, then it is wrong that the component does not directly affect the source of the input. It does, but as Box and Draper (1987) stated, "all models are wrong, but some are useful." An input to a model is essentially a statement about causation. The input causes behavior and is not caused by behavior. Of course, if the model is put into a feedback loop, then the input is affected by the output through the environment.

20. Sigmund, *Exact Thinking*, pp. 17–18.

21. Hitchcock, "What Russell Got Right," p. 54.

22. Churchland, *Touching a Nerve*, p. 178.

23. Churchland, *Touching a Nerve*, p. 180.

24. Dennett, *Intuition Pumps*, p. 358.

25. See Laplace (1901). In 2008, David Wolpert used a diagonalization technique to prove that Laplace's demon cannot exist (Wolpert, 2008). His proof relies on the observation that such a demon, were it to exist, would have to exist in the very physical world that it predicts. This results in a self-referentiality that yields contradictions, not unlike Turing's undecidability and Gödel's incompleteness theorems.

26. Also, in chapter 8, I referred to a mathematical result that I reported in 2016 (Lee, 2016), that shows that nondeterminism is unavoidable in any model of the physical world that includes both discrete and continuous behaviors if the model is rich enough to encompass Newton's laws.

27. Hawking, "Gödel."

28. In the most common interpretation, the Copenhagen interpretation, the wave function provides a probability density function for a random experiment that is performed when an observer makes an observation. This random experiment, often called the "collapse of the wave function," turns probability into certainty in a random way.

29. Popper, *Logic of Scientific Discovery.*

30. Pearl and Mackenzie, *Book of Why*, p. 362.

CHAPTER 12

1. Thelen, "Grounded in the World," p. 7.

2. Quisquater et al., "Zero-Knowledge Protocols."

3. Goldwasser et al., "Interactive Proof Systems (Extended Abstract)"; Goldwasser et al., "Interactive Proof Systems."

4. Goldwasser and Micali, "Probabilistic Encryption."

5. Zero-knowledge proofs were a first instance of a more general idea, interactive proofs, which, like the RCTs of the previous chapter, bring randomness and interaction together. The idea of interactive proofs was developed independently by László Babai, of Eőtvős University in Budapest and the University of Chicago (Babai, 1985). An interactive proof can be thought of as a game with two players, a prover (named Merlin by Babai) and a verifier (named Arthur by Babai). The verifier, Arthur, has limited ability to compute. Specifically, Arthur is assumed to be able to perform only computations that can be completed in a reasonable amount of time on a modern sequential computer, whereas the prover, Merlin, is allowed to perform more difficult computations. A reasonable amount of time is defined to be a time bounded by some polynomial function of the size of the statement being proved. In the story above, Shah Fi is the prover (Merlin) and Mick Ali is the verifier (Arthur).

6. Philippe et al., "Object-Comparison System."

7. Mathematicians are familiar with transfinite state structures in random systems. A class of such models are known as Markov processes, named after the Russian mathematician Andrey Markov. A simple and powerful example of a Markov process is a Wiener process, named after the same Norbert Wiener of the antiaircraft

guns that we encountered in chapter 5. A Wiener process is a continuous random walk, where you start at some point and wander at random in a space and time continuum. In physics, Wiener processes are used to model diffusion, to solve the Schrödinger equation, and to model cosmological inflation, for example. In engineering, they are used to model noise in electronics and disturbances in statistical mechanics and control theory. Wiener processes are also prominent in the mathematical theory of finance. They also play a role in the modeling of nerve behavior in biology (Ditlevsen and Samson, 2013).

8. Popper, *Logic of Scientific Discovery*.

9. The dashed-line notation is due to David Harel, who used it in a visual language that he invented called "StateCharts" (Harel, 1987).

10. This style of concurrent composition of beings is called "synchronous composition." From the Latin for "same time," "synchronous" reflects the fact that the two beings change states simultaneously.

11. I feel compelled to speculate that there might be some useful analogy with the observer problem in quantum physics. Perhaps the collapse of the wave function that is needed in the Copenhagen interpretation is just such a "peering into the soul" of the observed system. There might also be some useful analogy between two *Edward*s modeling each other and quantum entanglement. After *bang* but before *tick* or *tock*, these two *Edward*s are required to be in comparable states, poised to produce in the future the same *tick* or *tock*. Bell's theorem, named after the Irish physicist John Stewart Bell, uses quantum entanglement to rule out hidden variables as the source for experimentally observed randomness in quantum systems. Specifically, these experiments show that taking a measurement at one point in space can instantly affect the outcome of another experiment at a remote location, seemingly in violation of the speed-of-light limits on communication. Einstein called this property of quantum physics "spooky action at a distance." In a universe with two copies of *Edward* where one is modeling the other, there is an analogous violation because the modeler *Edward* must see the future of the *Edward* being modeled. Bell's theorem is often interpreted to mean that randomness in the physical world is real, but an equally explanatory resolution is that the world is actually deterministic in an extremely strong way, where every particle carries with it since its inception all the outcomes of all measurements that will ever be taken on it any time in the future. Since any measurement apparatus and any model must exist in the same universe, its particles too must carry with them their entire future. It will be this encoding of the future that will enable the modeler *Edward* to make "the right choice" at the time of its Little Bang. It sees its entire future all at once.

12. Lest any reader assume I am showing cultural bias by choosing Latino names when nondeterminism is involved, I must point out that I was called "Eduardo" when I was growing up in Puerto Rico and "Edward" throughout my adult life on the mainland. When I was young, my future was not determined, and I dallied through many possible career paths, including art, at which I proved to not be good enough. Much later in life, I became obsessed with determinism and became known

for advocating the use of deterministic models even for modeling nondeterministic systems. The choice of the name "Pablo" is, I hope, obvious. It is a bow to Pablo Picasso.

13. See Park (1980) and Milner (1989). Sangiorgi (2009) gives a nice overview of the historical development of this idea. He notes that essentially the same concept of bisimulation had also been developed in the fields of philosophical logic and set theory.

14. Matthew Peet, in a conversation we had in November 2018, was the first to suggest to me that the connection between Milner's way of modeling and zero-knowledge proofs was possibly stronger than I had realized. That conversation led to this model.

15. Philosophers use the term "intentionality" for "the power of minds to be about, to represent, or to stand for, things, properties, and states of affairs" outside the mind (Jacob, 2014). The Berkeley philosopher John Searle argues that intentionality is central to cognition (Searle, 1983). Intentionality is about models of the universe that we construct in our brains. Daniel Dennett suggests the less jargony term "aboutness" for intentionality (Dennett, 2013). The relationship between mental states and the things that those states are about is essentially a modeling relationship. In Milner's automata, the matching of states between the model and the automaton being modeled arguably provides a simple analogy to intentionality. Milner has shown us that aboutness is essential to modeling. Moreover, as we have seen, modeling works better when there is dialog, bidirectional interaction. Intentionality would likely not arise in a mind (or in a computer) that can only observe the world. It must also be able to affect the world.

CHAPTER 13

1. Copeland, *Essential Turing*, pp. 472–475.

2. Samuel Butler, *Erewhon*.

3. Personal communication, September 2018.

4. On May 25, 2016, the US General Accounting Office released a report stating that more than seventy percent of the US government's information technology budget is spent keeping old machines alive rather than "development, modernization, and enhancement." This amounts to about $65 billion per year, much of which is spent on outdated languages and hardware, some as much as fifty years old. As of 2016, the US Department of Defense still had eight-inch floppy disks in operation and the US Treasury Department still used programs written in assembly code.

5. Harari, *Homo Deus*, p. 21.

6. Rawls, *Theory of Justice*, p. 11.

7. Garner, *A Theory of Justice for Animals*.

8. An overview of the "datafication of health" is given by Ruckenstein and Schüll (2017).

9. Laland, *Darwin's Unfinished Symphony*, p. 263.

10. Bossert, "North Korea Behind WannaCry."

11. Husain, *Sentient Machine*.

12. Vincent, "Deepfakes for Dancing."

13. Rothman, "Is Seeing Still Believing?"

14. Dennett, *From Bacteria to Bach*.

15. The young activist Aviv Ovadya calls this an "infocalypse" (Warzel, 2018).

16. Gleick, *Information*.

17. Russell, *Human Compatible*.

18. Brooker, "'I Was Devastated.'"

19. Lee, *Super-Powers*.

20. Moor, "Machine Ethics."

21. Wiener, "Moral and Technical Consequences."

22. Armstrong, *Smarter Than Us*, p. 37.

23. Russell and Norvig, *Artificial Intelligence*.

24. Russell, "Learning Agents"; Ng and Russell, "Algorithms."

25. Hadfield-Menell et al., "Reinforcement Learning."

26. For an in-depth treatment of how divergent interests of humans may be combined to get safer AI behavior, see Russell (2019).

CHAPTER 14

1. Laland, *Darwin's Unfinished Symphony*, p. 6.

2. Laland, *Darwin's Unfinished Symphony*, p. 150.

3. Laland, *Darwin's Unfinished Symphony*, p. 234.

4. Turing, "Computing Machinery and Intelligence."

5. Quammen, *Tangled Tree*.

6. Quammen, *Tangled Tree*.

7. Zimmer, "Hunting Fossil Viruses in Human DNA."

8. Kelly, *Inevitable*.

9. Dawkins, *Selfish Gene*.

10. Dennett, *From Bacteria to Bach and Back*.

11. Lee, *Plato and the Nerd*, chapter 6.

12. Hillis, "Dawn of Entanglement."

13. Dennett, *From Bacteria to Bach*.

14. Dennett, *From Bacteria to Bach*.

15. Lee, *Plato and the Nerd*, chapter 9.

16. For a beautifully illustrated book documenting the effect of humans on the transformation of animals, see van Grouw (2008).

17. For examples of rapid Darwinian evolution of animals and plants in urban landscapes, see Schilthuizen (2018).

18. Dennett, *From Bacteria to Bach*.

19. Dennett, *From Bacteria to Bach*.

20. Lee, *Plato and the Nerd*, chapter 9.

21. Dennett, *From Bacteria to Bach*.

22. Dennett, *From Bacteria to Bach*.

23. Dennett, *From Bacteria to Bach*.

24. Smith and Douglas, *Biology of Symbiosis*.

25. Margulis and Sagan, *Acquiring Genomes*, pp. 14–15.

26. Margulis and Sagan, *Acquiring Genomes*, p. 141.

27. Mayr, *What Evolution Is*, p. 48.

28. Dyson, *Turing's Cathedral*, p. 311.

29. Margulis and Sagan, *Acquiring Genomes*, pp. 11–12.

30. Wilson et al., "Simple Programs."

31. Miller and Thomson, "Cartesian Genetic Programming."

32. Fitzpatrick et al., "Towards Long-lived Robot Genes."

33. Hofstadter, "Can Inspiration Be Mechanized?"

34. Pinker, *Blank Slate*, p. 417.

BIBLIOGRAPHY

Aguilar, Wendy, Guillermo Santamará-Bonfil, Tom Froese, and Carlos Gershenson. "The Past, Present, and Future of Artificial Life." *Frontiers in Robotics and AI* 1, no. 8 (October 2014): pp. 1–15. doi:10.3389/frobt.2014.00008.

Akkaya, Ilge, Daniel J. Fremont, Rafael Valle, Alexandre Donzé, Edward Lee, and Sanjit A. Seshia. "Control Improvisation with Probabilistic Temporal Specifications." In *Internet-of-Things Design and Implementation (IoTDI)*. IEEE, April 2016. doi:10.1109/IoTDI.2015.33.

Alcock, Joe, Carlo C. Maley, and C. Athena Aktipis. "Is Eating Behavior Manipulated by the Gastrointestinal Microbiota? Evolutionary Pressures and Potential Mechanisms." *BioEssays* 36, no. 10 (2014): pp. 940–949. doi:10.1002/bies.201400071.

Armstrong, Stuart. *Smarter Than Us: The Rise of Machine Intelligence*. Berkeley, CA: Machine Intelligence Research Institute, 2014.

Awad, Edmond, Sohan Dsouza, Richard Kim, Jonathan Schulz, Joseph Henrich, Azim Shariff, Jean-François Bonnefon, and Iyad Rahwan. "The Moral Machine Experiment." *Nature* (2018). doi:10.1038/s41586-018-0637-6.

Babai, László. "Trading Group Theory for Randomness." In *Symposium on Theory of Computing (STOC)*, pp. 421–429. New York: ACM, 1985. doi:10.1145/22145.22192.

Baraniuk, Chris. "The 'Creepy Facebook AI' Story That Captivated the Media." *BBC News*, August 2017. https://www.bbc.com/news/technology-40790258.

Barrat, James. *Our Final Invention: Artificial Intelligence and the End of the Human Era*. New York: St. Martin's Press, 2013.

Barry, John R., Edward A. Lee, and David G. Messerschmitt. *Digital Communication*. Third edition. New York: Springer Science + Business Media, LLC, 2004.

Bartneck, Christoph, Chioke Rosalia, Rutger Menges, and Inèz Deckers. "Robot Abuse—A Limitation of the Media Equation." In *Interac Workshop on Abuse*. 2005. https://www.bartneck.de/publications/2005/robotAbuse/bartneckInteract2005.pdf.

Benveniste, Albert, and Gérard Berry. "The Synchronous Approach to Reactive and Real-Time Systems." *Proceedings of the IEEE* 79, no. 9 (1991): pp. 1270–1282.

Berry, Gérard. *The Constructive Semantics of Pure Esterel*. Draft Version 3. Unpublished, 1999. http://www-sop.inria.fr/meije/esterel/doc/main-papers.html.

Black, H. S. "Stabilized Feed-back Amplifiers." *Electrical Engineering* 53 (1934): pp. 114–120.

Bongard, Josh, Victor Zykov, and Hod Lipson. "Resilient Machines Through Continuous Self-Modeling." *Science*, 2006, pp. 1118–1121.

Bossert, Thomas P. "It's Official: North Korea Is Behind WannaCry." *The Wall Street Journal*, December 2017.

Bostrom, Nick. "A History of Transhumanist Thought." *Journal of Evolution and Technology* 14, no. 1 (2005). https://nickbostrom.com/papers/history.pdf.

Bostrom, Nick. *Superintelligence: Paths, Dangers, Strategies*. Oxford, UK: Oxford University Press, 2014.

Box, George E. P., and Norman R. Draper. *Empirical Model-Building and Response Surfaces*. Wiley Series in Probability and Statistics. Hoboken, NJ: Wiley, 1987.

Bratsberg, Bernt, and Ole Rogeberg. "Flynn Effect and Its Reversal Are Both Environmentally Caused." *Proceedings of the National Academy of Sciences of the United States of America*, 2018. doi:10.1073/pnas.1718793115.

Brooker, Katrina. "'I Was Devastated': Tim Berners-Lee, The Man Who Created the World Wide Web, Has Some Regrets." *Vanity Fair*, July 2018. https://www.vanityfair.com/news/2018/07/the-man-who-created-the-world-wide-web-has-some-regrets.

Brooks, Rodney A. "Artificial Life and Real Robots." In *Toward a Practice of Autonomous Systems: Proceedings of the First European Conference on Artificial Life*, pp. 3–10. Cambridge, MA: MIT Press, 1992.

Brown, Dan. *Origin: A Novel*. New York: Doubleday, 2017.

Brscic, Drazen, Hiroyuki Kidokoro, Yoshitaka Suehiro, and Takayuki Kanda. "Escaping from Children's Abuse of Social Robots." In *International Conference on Human-Robot Interaction (HRI)*. ACM/IEEE, March 2015.

Bryson, Arthur E., Walter F. Denham, Frank J. Carroll, and Kinya Mikami. "A Steepest-Ascent Method for Solving Optimum Programming Problems." *Journal of Applied Mechanics* 29, no. 2 (1961): pp. 247–257. doi:10.1115/1.3640537.

Butler, Samuel. *Erewhon: or Over the Range*. London: Trübner, second edition, 1872). Available from the Gutenberg Project at http://www.gutenberg.org/ebooks/1906.

Carmena, José M., Mikhail A. Lebedev, Roy E. Crist, Joseph E. O'Doherty, David M. Santucci, Dragan F. Dimitrov, Parag G. Patil, Craig S. Henriquez, and Miguel A. L. Nicolelis. "Learning to Control a Brain-Machine Interface for Reaching and Grasping by Primates." *PLoS Biol* 1, no. 2 (2003). doi:10.1371/journal.pbio.0000042.

Caruana, Rich, Yin Lou, Paul Koch, Marc Sturm, Johannes Gehrke, and Noémie Elhadad. "Intelligible Models for HealthCare: Predicting Pneumonia Risk and Hospital 30-Day Readmission." *ACM SIGKDD Conference on Knowledge Discovery and Data Mining (KDD)*, pp. 1721–1730. 2015. doi:10.1145/2783258.2788613.

Casselman, Anne. "Identical Twins' Genes Are Not Identical." *Scientific American*. https://www.scientificamerican.com/article/identical-twins-genes-are-not-identical/.

Chaitin, Gregory. "How Real Are Real Numbers?" *ArXiv* arXiv:math/0411418v3 [math.HO] (2004). https://arxiv.org/abs/math/0411418v3.

Chaitin, Gregory. *Meta Math!: The Quest for Omega*. New York: Vintage Books, 2005.

Chalmers, David J. *The Conscious Mind: In Search of a Fundamental Theory*. Oxford, UK: Oxford University Press, 1996.

Chesterton, G. K. *G. F. Watts*. New York: Cosimo, 1904, reprinted 2007.

Churchland, Patricia. *Touching a Nerve: Our Brains, Our Selves*. New York: W. W. Norton, 2013.

Clark, Andy. *Supersizing the Mind: Embodiment, Action, and Cognitive Extension*. Oxford, UK: Oxford University Press, 2008.

Clark, Andy, and David Chalmers. "The Extended Mind." *Analysis* 58, no. 1 (1998): 7–19. doi:10.1111/1467-8284.00096.

Clarke, Arthur C. *2001: A Space Odyssey*. London: Hutchinson/Company, 1968.

Cooper, Gregory F., Constantin F. Aliferis, Richard Ambrosino, John Aronis, Bruce G. Buchanan, Richard Caruana, Michael J. Fine, Clark Glymour, Geoffrey Gordon, Barbara H. Hanusa, Janine E. Janosky, Christopher Meek, Tom Mitchell, Thomas Richardson, Peter Spirtes, "An Evaluation of Machine-Learning Methods for Predicting Pneumonia Mortality." *Artificial Intelligence in Medicine* 9 (1997): pp. 107–138.

Copeland, B. Jack. "The Church-Turing Thesis." *The Stanford Encyclopedia of Philosophy*, Winter 2017. https://plato.stanford.edu/archives/win2017/entries/church-turing.

Copeland, B. Jack. *The Essential Turing: Seminal Writings in Computing, Logic, Philosophy, Artificial Intelligence, and Artificial Life Plus The Secrets of Enigma*. Oxford, UK: Oxford University Press, 2004.

Copley, Bridget, and Fabienne Martin. *Causation in Grammatical Structures*. Oxford, UK: Oxford University Press, 2014.

Danziger, Shai, Jonathan Levav, and Liora Avnaim-Pesso. "Extraneous Factors in Judicial Decisions." *Proceedings of the National Academy of Sciences of the United States of America* 108, no. 17 (2011): 6889–6892. doi:10.1073/pnas.1018033108.

Davies, Alex. "A Sleeping Tesla Driver Highlights Autopilot's Biggest Flaw." *WIRED*, December 2018. https://www.wired.com/story/tesla-sleeping-driver-dui-arrest-autopilot/.

Dawkins, Richard. *The Blind Watchmaker: Why the Evidence of Evolution Reveals a Universe Without Design*. New York: Norton, 1987.

Dawkins, Richard. *The Selfish Gene*. Oxford, UK: Oxford University Press, 1976.

Dennett, Daniel C. *Consciousness Explained*. New York: Back Bay Books, 1991.

Dennett, Daniel C. *Elbow Room: The Varieties of Free Will Worth Wanting*. Cambridge, MA: MIT Press, 1984, 2015.

Dennett, Daniel C. *From Bacteria to Bach and Back: The Evolution of Minds*. New York: W. W. Norton, 2017.

Dennett, Daniel C. *Intuition Pumps and Other Tools for Thinking*. New York: W. W. Norton, 2013.

Dhar, Abhishek. "Nonuniqueness in the Solutions of Newton's Equation of Motion." *American Journal of Physics* 61, no. 1 (1993): pp. 58–61. doi:10.1119/1.17411.

Dickinson, Emily. *The Complete Poems of Emily Dickinson*, edited by Thomas H. Johnson. Boston, Toronto: Little Brown/Company, 1890.

Ditlevsen, Susanne, and Adeline Samson. "Introduction to Stochastic Models in Biology." In *Stochastic Biomathematical Models: with Applications to Neuronal Modeling*, 3–35. New York: Springer, 2013. doi:10. 1007/978-3-642-32157-3_1.

Dodig-Crnkovic, Gordana. *Investigations into Information Semantics and Ethics of Computing*. vol. 33. Dissertations. Västerås, Sweden: Mälardalen University Press, 2006. http://mdh.diva-portal.org/smash/get/diva2:120541/FULLTEXT01.

Donzé, Alexandre, Raphael Valle, Ilge Akkaya, Sophie Libkind, Sanjit A. Seshia, and David Wessel. "Machine Improvisation with Formal Specifications." In *International Computer Music Conference (ICMC)*, pp. 1277–1284. September 2014.

Doyle, Bob. *Free Will: The Scandal in Philosophy*. Cambridge, MA: I-Phi Press, 2011.

Dreyfus, Hubert L., and Stuart E. Dreyfus. "From Socrates to Expert Systems: The Limits of Calculative Rationality." *Technology in Society* 6, no. 3 (1984): pp. 217–233. doi:10.1016/0160-791X(84)90034-4.

Dreyfus, Hubert L., and Stuart E. Dreyfus. *Mind Over Machine: The Power of Human Intuition and Expertise in the Era of the Computer*. New York: Free Press, 1986. doi:10.1016/0160-791X(84)90034-4.

Dreyfus, Stuart E. "Artificial Neural Networks, Back Propagation, and the Kelley-Bryson Gradient Procedure." *Journal on Guidance, Control, and Dynamics* 13, no. 5 (1990): pp. 926–928. doi:10.2514/3.25422.

Dyson, George. *Darwin Among the Machines: The Evolution of Global Intelligence*. New York: Basic Books, 1997.

Dyson, George. *Turing's Cathedral: The Origins of the Digital Universe*. New York: Pantheon Books, 2012.

Earman, John. *A Primer on Determinism*. vol. 32. The University of Ontario Series in Philosophy of Science. Dordrecht, Holland: D. Reidel Publishing Company, 1986.

Emmeche, Claus. *The Garden in the Machine*. Princeton, NJ: Princeton University Press, 1994.

England, Jeremy L. "Statistical Physics of Self-replication." *The Journal of Chemical Physics* 139, no. 121923 (2013): pp. 1–8.

Fitzpatrick, Paul, Giorgio Metta, and Lorenzo Natale. "Towards Long-lived Robot Genes." *Robotics and Autonomous Systems* 56, no. 1 (2008): pp. 29–45. doi:10.1016/j.robot.2007.09.014.

Ford, Martin. *Rise of the Robots: Technology and the Threat of a Jobless Future*. New York: Basic Books, 2015.

Fremont, Daniel J., Alexandre Donzé, Sanjit A. Seshia, and David Wessel. "Control Improvisation." In *Foundations of Software Technology and Theoretical Computer Science (FSTTCS)*, pp. 463–474. 2015. https://arxiv.org/abs/1704.06319.

Fried, Iltzhak, Roy Mukamel, and Gabriel Kreiman. "Internally Generated Preactivation of Single Neurons in Human Medial Frontal Cortex Predicts Volition." *Neuron* 69, no. 3 (2011): pp. 548–562. doi:10.1016/j.neuron.2010.11.045.

Garner, Robert. *A Theory of Justice for Animals: Animal Rights in a Nonideal World*. Oxford, UK: Oxford University Press, 2013.

Gaudiano, Anora M. "Here's One Key Factor That Amplified the 1987 Stock-Market Crash." *Market Watch*, October 1987. https://www.marketwatch.com/story/heres-one-key-factor-that-amplified-the-1987-stock-market-crash-2017-10-16.

Giles, Martin. "The GANfather: The Man Who's Given Machines the Gift of Imagination." *MIT Technology Review* 121, no. 2 (2018): pp. 48–53.

Gleick, James. *Genius: The Life and Science of Richard Feynman*. New York: Vintage Books, 1993.

Gleick, James. *The Information: A History, A Theory, A Flood*. New York: Pantheon Books, 2011.

Godfrey-Smith, Peter. *Other Minds: The Octopus, the Sea, and the Deep Origins of Consciousness*. New York: Farrar, Straus/Giroux, 2016.

Goldberg, Ken. "The Robot-Human Alliance." *Wall Street Journal* Op Ed (2017).

Goldin, Dina, and Peter Wegner. "The Church-Turing Thesis: Breaking the Myth." In *New Computational Paradigms*, vol. LNCS 3526, pp. 152–168. New York: Springer, 2005.

Goldwasser, Shafi, and Silvio Micali. "Probabilistic Encryption." *Journal of Computer and System Sciences (JCSS)* 28, no. 2 (1984): pp. 270–299.

Goldwasser, Shafi, Silvio Micali, and Charles Rackoff. "The Knowledge Complexity of Interactive Proof Systems (Extended Abstract)." In *Symposium on Theory of Computing (STOC)*, 291–304. New York: ACM, 1985.

Goldwasser, Shafi, Silvio Micali, and Charles Rackoff. "The Knowledge Complexity of Interactive Proof Systems." *SIAM Journal on Computing* 18, no. 1 (1989): pp. 186–208. doi:10.1137/0218012.

Goldwasser, Shafi, Silvio Micali, and Charles Rackoff. "The Knowledge Complexity of Interactive Proof Systems (Extended Abstract)." In *Symposium on Theory of Computing (STOC)*, 291–304. ACM, 1985.

Good, Irving John. "Speculations Concerning the First Ultraintelligent Machine." *Advances in Computers* 6 (1966): pp. 31–88.

Gribbin, John. *Alone in the Universe: Why Our Planet Is Unique*. Hoboken, NJ: John Wiley & Sons, 2011.

Grüsser, Otto-Joachim. "On the History of the Ideas of Efference Copy and Reafference." In *Essays in the History of Physiological Sciences: Proceedings of a Symposium Held at the University Louis Pasteur Strasbourg, on March 2627th, 1993*, 33: pp. 35–56. London: The Wellcome Institute Series in the History of Medicine. Clio Medica, 1995.

Hadfield-Menell, Dylan, Stuart J Russell, Pieter Abbeel, and Anca Dragan. "Cooperative Inverse Reinforcement Learning." In *Advances in Neural Information Processing Systems 29*, edited by D. D. Lee, M. Sugiyama, U. V. Luxburg, I. Guyon, and R. Garnett, pp. 3909–3917. Red Hook, NY: Curran Associates, Inc., 2016. http://papers.nips.cc/paper/6420-cooperative-inverse-reinforcement-learning.pdf.

Haggard, Patrick. "Decision Time for Free Will." *Neuron* 69, no. 3 (2011): pp. 404–406. doi:10.1016/j.neuron.2011.01.028.

Handwerk, Brian. "Your Gut Bacteria May Be Controlling Your Appetite." *Smithonian.com*. https:///www.smithsonianmag.com/science-nature/gut-bacteria-may-be-controlling-your-appetite.

Harari, Yuval Noah. *Homo Deus: A Brief History of Tomorrow*. New York: HarperCollins, 2017.

Harari, Yuval Noah. *21 Lessons for the 21st Century*. New York: Penguin Random House, 2018.

Harel, David. "Statecharts: A Visual Formalism for Complex Systems." *Science of Computer Programming* 8, no. 3 (1987): pp. 231–274.

Harris, Sam. *Free Will*. New York: Free Press, 2012.

Hart, Mathew. "Giant Wall of Lava Lamps Helps Protect 10% of Internet Traffic (Seriously)." Nerdist, November 2017. https://nerdist.com/wall-of-lava-lamps-protect-internet-traffic/.

Hartnett, Kevin. "Secret Link Uncovered Between Pure Math and Physics." *Quanta Magazine*, December 2017. https://www.quantamagazine.org/secret-link-uncovered-between-pure-math-and-physics-20171201/.

Haugeland, John. *Artificial Intelligence: The Very Idea*. Cambridge, MA: MIT Press, 1985.

Hawking, Stephen. "Gödel and the End of the Universe." *Stephen Hawking Public Lectures*, 2002. http://www.hawking.org.uk/godel-and-the-end-of-physics.html.

Haynes, John-Dylan. "Decoding and Predicting Intentions." *Annals of the New York Academy of Science* 1224 (2011): pp. 9–21. doi:10.1111/j.1749-6632.2011.05994.x.

Hillis, W. Daniel. "Introduction: The Dawn of Entanglement." In *Is the Internet Changing the Way You Think?: The Net's Impact on Our Minds and Future*. pp. 1–14, New York: Harper Collins, 2011.

Hitchcock, Christopher. "What Russell Got Right." In *Causation, Physics, and the Constitution of Reality*, pp. 45–65. Oxford, UK: Clarendon Press, 2007.

Hobbes, Thomas. *Leviathan: or the Matter, Forme, & Power of a Common-wealth Ecclesiasticall and Civil*. London: Andrew Crooke, 1651. https://socialsciences.mcmaster.ca/econ/ugcm/3ll3/hobbes/Leviathan.pdf.

Hofstadter, Douglas. "Can Inspiration Be Mechanized?" *Scientific American* 247, no. 3 (1982): pp. 18–34.

Hofstadter, Douglas. *Gödel, Escher, and Bach: An Eternal Golden Braid*. New York: Basic Books, 1979.

Hofstadter, Douglas. *I Am a Strange Loop*. New York: Basic Books, 2007.

Husain, Amir. *The Sentient Machine: The Coming Age of Artificial Intelligence*. New York: Scribner, 2017.

Hutter, Marcus. *Universal Artificial Intelligence: Sequential Decisions Based on Algorithmic Probability*. Berlin: Springer, 2004.

Huxley, Julian S. *Religion Without Revelation*. New York: Harper & Brothers Publishers, 1927.

Jacob, Pierre. "Intentionality." *Stanford Encyclopedia of Philosophy*, October 2014. http://plato.stanford.edu/archives/win2014/entries/intentionality/.

Kahneman, Daniel. *Thinking Fast and Slow*. New York: Farrar, Straus/Giroux, 2011.

Kastrenakes, Jacob. "Microsoft Made a Chatbot That Tweets Like a Teen." *The Verge*, 2016. https://www.theverge.com/2016/3/23/11290200/tay-ai-chatbot-released-microsoft.

Kauffman, Stuart. *The Origins of Order: Self-Organization and Selection in Evolution*. Oxford, UK: Oxford University Press, 1993.

Kelley, Henry J. "Gradient Theory of Optimal Flight Paths." *ARS Journal* 30, no. 10 (1960): 947–954. doi:10.2514/8.5282.

Kelly, Kevin. *The Inevitable: Understanding the 12 Technological Forces That Will Shape Our Future*. New York: Penguin Books, 2016.

Kelly, Kevin. *What Technology Wants*. New York: Penguin Books, 2010.

Kinniment, David J. *Synchronization and Arbitration in Digital Systems*. New York: John Wiley & Sons, 2007.

Kosinski, Michal, David Stillwell, and Thore Graepel. "Private Traits and Attributes Are Predictable from Digital Records of Human Behavior." Published ahead of print. *Proceedings of the National Academy of Sciences of the United States of America*, 2013. doi:10.1073/pnas.1218772110.

Kurzweil, Raymond. *The Age of Intelligent Machines*. Cambridge, MA: MIT Press, 1990.

Lake, Brenden M., Ruslan Salakhutdinov, and Joshua B. Tenenbaum. "Human-Level Concept Learning Through Probabilistic Program Induction." *Science* 350, no. 6266 (2015): pp. 1332–1338. doi:10.1126/science.aab3050.

Laland, Kevin N. *Darwin's Unfinished Symphony: How Culture Made the Human Mind*. Princeton, NJ: Princeton University Press, 2017.

Langton, Christopher G. *A New Definition of Artificial Life*. Unpublished Report. 1998. http://scifunam.fisica.unam.mx/mir/langton.pdf.

Langton, Christopher G., ed. *Artificial Life: Proceedings of an Interdisciplinary Workshop on the Synthesis and Simulation of Living Systems*. Boston: Addison-Wesley, 1988.

Laplace, Pierre-Simon. *A Philosophical Essay on Probabilities*. Translated from the sixth French edition by F. W. Truscott and F. L. Emory. Hoboken, NJ: John Wiley & Sons, 1901.

Lee, Bernard S. "Effects of Delayed Speech Feedback." *Journal of the Acoustical Society of America* 22, no. 6 (1950): pp. 824–826. doi:10. 1121/1.1906696.

Lee, Edward A. "Constructive Models of Discrete and Continuous Physical Phenomena." *IEEE Access* 2, no. 1 (2014): pp. 1–25. doi:10.1109/ACCESS.2014.2345759.

Lee, Edward A. "Fundamental Limits of Cyber-Physical Systems Modeling." *ACM Transactions on Cyber-Physical Systems* 1, no. 1 (2016): pp. 3:1–3:26. doi:10.1145/2912149.

Lee, Edward A., and David G. Messerschmitt. *Digital Communication*. Boston: Kluwer Academic Publishers, 1988.

Lee, Edward Ashford. *Plato and the Nerd: The Creative Partnership of Humans and Technology*. Cambridge, MA: MIT Press, 2017.

Lee, Kai-Fu. *Super-Powers: China, Silicon Valley, and the New World Order*. New York: Houghton Mifflin Harcourt Publishing Company, 2018.

Legg, Shane, and Marcus Hutter. "A Universal Measure of Intelligence for Artificial Agents." In *International Joint Conference on Artificial Intelligence (IJCAI)*, pp. 1509–1510. Mahwah NJ: Lawrence Erlbaum, 2005. http://www.ijcai.org/papers /post-0042.pdf.

Leland, Hayne E., and Mark Rubinstein. "The Evolution of Portfolio Insurance." In *Dynamic Hedging: A Guide to Portfolio Insurance.* pp. 1–7, New York: John Wiley & Sons, 1988.

Libet, Benjamin. "Unconscious Cerebral Initiative and the Role of Conscious Will in Voluntary Action." *Behavioral and Brain Sciences* 8, no. 4 (1985): pp. 529–539. doi:10.1017/S0140525X00044903.

Libet, Benjamin, Curtis A. Gleason, Elwood W. Wright, and Dennis K. Pearl. "Time of Conscious Intention to Act in Relation to Onset of Cerebral Activity (Readiness-Potential): The Unconscious Initiation of a Freely Voluntary Act." *Brain* 106, no. 3 (1983): pp. 623–642. doi:10.1093/brain/106.3.623.

Lichtman, Jeff W. "Can the Brain's Structure Reveal Its Function, Theoretically Speaking?" In *Theoretically Speaking Series.* Invited talk. Simons Institute for the Theory of Computing, 2018. https://simons.berkeley.edu/events/theoretically -speaking-jeff-lichtman.

Lichtman, Jeff W., Hanspeter Pfister, and Nir Shavit. "The Big Data Challenges of Connectomics." *Nature Neuroscience* 17 (2014): pp. 1448–1454. doi:10.1038/nn.3837.

Lucas, John Randolph. "Minds, Machines, and Gödel." *Philosophy* 36, no. 137 (1961): pp. 112–127.

MacKay, Donald G. "Metamorphosis of a Critical Interval: Age-Linked Changes in the Delay in Auditory Feedback that Produces Maximal Disruption of Speech." *Journal of the Acoustical Society of America* 43, no. 4 (2005): pp. 811–821. doi:10.1121/1.1910900.

Margulis, Lynn, and Dorion Sagan. *Acquiring Genomes: A Theory of the Origins of Species.* New York: Basic Books, 2002.

Marino, Leonard R. "General Theory of Metastable Operation." *IEEE Transactions on Computers* C-30, no. 2 (1981): pp. 107–115.

Maturana, Humberto, and Francisco Varela. *Autopoiesis and Cognition: the Realization of the Living.* Dortrecht, Boston, London: D. Reidel Publishing Company, 1980.

Mayr, Ernst. *What Evolution Is.* New York: Basic Books, 2001.

McLuhan, Marshall. *The Gutenberg Galaxy: The Making of Typographic Man.* Toronto: University of Toronto Press, 1962.

McLuhan, Marshall. *Understanding Media: The Extensions of Man.* New York: McGraw Hill, 1964.

Mendler, Michael, Thomas R. Shiple, and Gérard Berry. "Constructive Boolean Circuits and the Exactness of Timed Ternary Simulation." *Formal Methods in System Design* 40, no. 3 (2012): pp. 283–329. doi:10.1007/s10703-012-0144-6.

Miller, Julian F., and Peter Thomson. "Cartesian Genetic Programming." In *European Conference on Genetic Programming*, vol. LNCS vol. 10802, pp. 121–132. New York: Springer, 2000.

Milner, Robin. *Communication and Concurrency*. Englewood Cliffs, NJ: Prentice Hall, 1989.

Mitchell, Tom M. *Machine Learning*. New York: McGraw Hill, 1997.

Moor, James H. "The Nature, Importance, and Difficulty of Machine Ethics." *IEEE Intelligent Systems* 21, no. 4 (2006): pp. 18–21. doi:10.109/MIS.2006.80.

Muller, Richard A. *Now: The Physics of Time*. W. W. Norton, 2016.

Ng, Andrew, and Stuart Russell. "Algorithms for Inverse Reinforcement Learning." in *Proc. 17th International Conf. on Machine Learning*, pp. 663–670. San Francisco, CA: Morgan Kaufman, 2000.

Nietzsche, Friedrich. *The Will to Power*. Translated by Walter Kaufmann and R. J. Hollingdale. New York: Vintage Books, 1886–87, 1967 translation.

Norton, John D. "Causation as Folk Science." In *Causation, Physics, and the Constitution of Reality*, pp. 11–44. Oxford, UK: Clarendon Press, 2007.

Parfit, Derek. *Reasons and Persons*. New York: Oxford University Press, 1984.

Park, David. "Concurrency and Automata on Infinite Sequences." In *Theoretical Computer Science*, vol. LNCS 104. Berlin, Heidelberg: Springer, 1980. doi:10.1007/BFb0017309.

Parker, Andrew. *In the Blink of an Eye: How Vision Sparked the Big Bang of Evolution*. New York: Perseus Pub, 2003.

Pearl, Judea, and Dana Mackenzie. *The Book of Why: The New Science of Cause and Effect*. New York: Basic Books, 2018.

Penrose, Roger. *The Emperor's New Mind: Concerning Computers, Minds and The Laws of Physics*. Oxford, UK: Oxford University Press, 1989.

Pfeifer, Rolf, and Josh Bongard. *How the Body Shapes the Way We Think: A New View of Intelligence*. Cambridge, MA: MIT Press, 2007.

Philippe, Sébastien, Robert J. Goldston, Alexander Glaser, and Francesco d'Errico. "A Physical Zero-Knowledge Object-Comparison System for Nuclear Warhead Verification." *Nature Communications* 7 (2016): pp. 1–7. doi:10.1038/ncomms12890.

Pinker, Steven. *The Blank Slate: The Modern Denial of Human Nature*. New York: Viking, 2002/2016.

Pinker, Steven. *Enlightenment Now: The Case for Reason, Science, Humanism, and Progress*. New York: Penguin Books, 2018.

Pollock, Cassandra, and Alex Samuels. "Hysteria Over Jade Helm Exercise in Texas Was Fueled by Russians, Former CIA Director Says." *Texas Tribune*, May 2018.

https://www.texastribune.org/2018/05/03/hysteria-over-jade-helm-exercise-texas-was-fueled-russians-former-cia-/.

Popper, Karl. *The Logic of Scientific Discovery*. London: Hutchinson & Co., 1959.

Putnam, Hilary. "Psychological Predicates." In *Art, Mind, and Religion*, pp. 37–48. Pittsburgh: University of Pittsburgh Press, 1967.

Quammen, David. *The Tangled Tree: A Radical New History of Life*. New York: Simon & Schuster, 2018.

Quisquater, Jean-Jacques (with Myriam, Mureil, and Michaël), Louis C. Guillou (with Marie Annick, and Gäıd, Genolé, and Soazig), and with Tom Berson (for the English translation). "How to Explain Zero-Knowledge Protocols to Your Children." In *Advances in Cryptology (CRYPTO)*, pp. 628–631. New York: Springer, 1989.

Rawls, John. *A Theory of Justice*. Cambridge, MA: Harvard University Press, 1971.

Ribeiro, Marco Túlio, Sameer Singh, and Carlos Guestrin. "Why Should I Trust You? Explaining the Predictions of Any Classifier." In *International Conference on Knowledge Discovery and Data Mining*, pp. 1135–1144. New York: ACM, 2016. doi:10.1145/2939672.2939778.

Rogers, Deborah S., and Paul R. Ehrlich. "Natural Selection and Cultural Rates of Change." *Proceedings of the National Academy of Sciences of the United States of America* 105, no. 9 (2008): pp. 3416–3420.

Rosenblueth, Arturo, Norbert Wiener, and Julian Bigelow. "Behavior, Purpose, and Teleology." *Philosophy of Science* 10, no. 1 (1943): pp. 18–24.

Rothman, Joshua. "In the Age of A.I., Is Seeing Still Believing?" *New Yorker*, November 2018. https://www.newyorker.com/magazine/2018/11/12/in-the-age-of-ai-is-seeing-still-believing.

Rovelli, Carlo. *The Order of Time*. New York: Riverhead Books, 2018.

Ruckenstein, Minna, and Natasha Dow Schüll. "The Datafication of Health." *Annual Review of Anthropology* 46 (2017): pp. 261–278. doi:10.1146/annurev-anthro-102116-041244.

Rumelhart, David E., Geoffrey E. Hinton, and Ronald J. Williams. "Learning Representations by Back-propagating Errors." *Nature* 323 (1986): pp. 533–536.

Russell, Bertrand. "On the Notion of Cause." *Proceedings of the Aristotelian Society* 13 (1913): pp. 1–26.

Russell, Stuart. *Human Compatible: Artificial Intelligence and the Problem of Control*. New York: Viking, 2019.

Russell, Stuart. "Learning Agents for Uncertain Environments (Extended Abstract)." In *Computational Learning Theory (COLT)*, New York: ACM pp. 101–103. July 1998. doi:10.1145/279943.279964.

Russell, Stuart, and Peter Norvig. *Artificial Intelligence*. London: Pearson, 2010.

Sagan, Lynn. "On the Origin of Mitosing Cells." *Journal of Theoretical Biology* 14, no. 3 (1967): pp. 225–274. doi:10.1016/0022-5193(67)90079-3.

Sangiorgi, Davide. "On the Origins of Bisimulation and Coinduction." *ACM Transactions on Programming Languages and Systems* 31, no. 4 (2009): 15:1–15:41. doi:10.1145/1516507.1516510.

Sapolsky, Robert M. *Behave: The Biology of Humans at Our Best and Worst*. New York: Penguin Press, 2017.

Schilthuizen, Menno. *Darwin Comes to Town: How the Urban Jungle Drives Evolution*. New York: Picador, Macmillan Publishing Group, 2018.

Schrödinger, Erwin. *What Is Life: The Physical Aspect of the Living Cell*. Cambridge, UK: Cambridge University Press, 1944.

Searle, John R. *Intentionality: An Essay in the Philosophy of Mind*. Cambridge, UK: Cambridge University Press, 1983.

Shannon, Claude E. "A Mathematical Theory of Communication." Reprinted in 2001 with corrections from the *Bell System Technical Journal*, 1948. *ACM SIGMOBILE Mobile Computing and Communications Review* 5, no. 1 (1948): pp. 3–55. doi:10.1145/584091.584093.

Shapiro, Ehud. "A Mechanical Turing Machine: Blueprint for a Biomolecular Computer." *Interface Focus* 2, no. 4 (2012): pp. 497–503. doi:10.1098/rsfs.2011.0118.

Sigmund, Karl. *Exact Thinking in Demented Times: The Vienna Circle and the Epic Quest for the Foundations of Science*. New York: Basic Books, 2017.

Simonite, Tom. "When It Comes to Gorillas, Google Photos Remains Blind." *WIRED*, January 2018. https://www.wired.com/story/when-it-comes-to-gorillas-google-photos-remains-blind/.

Smith, David C., and Angela E. Douglas. *The Biology of Symbiosis*. London: Edward Arnold (Publishers) Ltd., 1987.

Sorensen, Eric, and Jeff Reinke. "Nissan Braking System Deactivating Itself." *IEN (Industrial Equipment News) Newsletter*, September 2018. https://www.ien.com/product-development/video/21024265/nissan-braking-system-deactivating-itself.

Stringer, Christopher. "Brain Size Has Increased for Most of Our Existence, So Why Has It Started to Diminish for the Past Few Thousand Years?" *Scientific American Mind* 25 (October 2014): pp. 74–74. doi:10.1038/scientificamericanmind1114-74b.

Taleb, Nassim Nicholas. *The Black Swan*. New York: Random House, 2010.

Tegmark, Max. *Life 3.0: Being Human in the Age of Artificial Intelligence*. New York: Alfred A. Knopf, 2017.

Thelen, Esther. "Grounded in the World: Developmental Origins of the Embodied Mind." *Infancy* 1, no. 1 (2000): pp. 3–28.

Tian, Xing, and David Poeppel. "Mental Imagery of Speech and Movement Implicates the Dynamics of Internal Forward Models." *Frontiers in Psychology*, 1, article 166 pp. 1–23 2010. doi:10.3389/fpsyg.2010.00166.

Turing, Alan. M. "Computing Machinery and Intelligence." *Mind* 59, no. 236 (1950): pp. 433–460. http://www.jstor.org/stable/2251299.

Turing, Alan M. "On Computable Numbers with an Application to the Entscheidungsproblem." *Proceedings of the London Mathematical Society* 42 (1936): pp. 230–265.

van Grouw, Katrina. *Unnatural Selection*. Princeton, NJ: Princeton University Press, 2008.

Vigna, Paul, and Michael J. Casey. *The Age of Cryptocurrency: How Bitcoin and Digital Money Are Challenging the Global Economic Order*. New York: St. Martin's Press, 2015.

Vincent, James. "Deepfakes for Dancing: You Can Now Use AI to Fake Those Dance Moves You Always Wanted." *The Verge*, August 2018. https://www.theverge.com/2018/8/26/17778792/deepfakes-video-dancing-ai-synthesis.

Vincent, James. "How Three French Students Used Borrowed Code to Put the First AI Portrait in Christie's." *The Verge*, October 2018. https://www.theverge.com/2018/10/23/18013190/ai-art-portrait-auction-christies-belamy-obvious-robbie-barrat-gans.

Vincent, James. "Lyrebird Claims It Can Recreate Any Voice Using Just One Minute of Sample Audio." *The Verge*, April 2017. https://www.theverge.com/2017/4/24/15406882/ai-voice-synthesis-copy-human-speech-lyrebird.

Vincent, James. "Twitter Taught Microsoft's AI Chatbot to Be a Racist Asshole in Less Than a Day." *The Verge*, March 2016. https://www.theverge.com/2016/3/24/11297050/tay-microsoft-chatbot-racist.

Vinge, Vernor. "The Coming Technological Singularity." *Whole Earth Review* Winter issue (1993).

von Neumann, John. "The General and Logical Theory of Automata." In *Hixon Symposium*, 1–41. Hafner Publishing, September 1951.

Wachter, Sandra, Brent Mittelstadt, and Luciano Floridi. "Why a Right to Explanation of Automated Decision-Making Does Not Exist in the General Data Protection Regulation." *International Data Privacy Law, Available at SSRN*, 2017. doi:10.2139/ssrn.2903469. https://ssrn.com/abstract=2903469.

Wang, Yilun, and Michal Kosinski. "Deep Neural Networks Are More Accurate than Humans at Detecting Sexual Orientation from Facial Images." *Journal of Personality and Social Psychology* 114, no. 2 (2018): pp. 246–257. doi:10.1037/pspa0000098.

Warzel, Charlie. "He Predicted The 2016 Fake News Crisis. Now He's Worried About An Information Apocalypse." *Buzz Feed*, 2018. https://www.buzzfeednews.com/article/charliewarzel/the-terrifying-future-of-fake-news.

Wegner, Peter. "Why Interaction Is More Powerful Than Algorithms." *Communications of the ACM* 40, no. 5 (1997): pp. 80–91. doi:10.1145/253769.253801.

Wegner, Peter, Farhad Arbab, Dina Goldin, Peter McBurney, Michael Luck, and Dave Roberson. "The Role of Agent Interaction in Models of Computing: Panelist Reviews." *Electronic Notes in Theoretical Computer Science* 141 (2005): pp. 181–198. https://eprints.soton.ac.uk/261913/1/finco05.pdf.

Weizenbaum, Joseph. "ELIZA—A Computer Program for the Study of Natural Language Communication Between Man and Machine." *Communications of the ACM* 9, no. 1 (1966): pp. 36–45. doi:10.1145/365153.365168.

Wheeler, John Archibald. "Hermann Weyl and the Unity of Knowledge." *American Scientist* 74 (1986): pp. 366–375. http://www.weylmann.com/wheeler.pdf.

Wiener, Norbert. *Cybernetics: Or Control and Communication in the Animal and the Machine*. Cambridge, MA: Librairie Hermann & Cie, Paris/MIT Press, 1948.

Wiener, Norbert. "Some Moral and Technical Consequences of Automation." *Science of Computer Programming* 131 pp. 1355–1358 (1960).

Wilson, Dennis G., Sylvain Cussat-Blanc, Hervé Luga, and Julian F Miller. "Evolving Simple Programs for Playing Atari Games." In *The Genetic and Evolutionary Computation Conference (GECCO)*. New York: ACM, pp. 229–236, June 2018. doi:10.1145/3205455.3205578.

Wittgenstein, Ludwig. *Tractatus Logico-Philosophicus*. Translated by Charles Kay Ogden. London: Routledge & Kegan Paul Ltd., 1922, 1960 translation.

Wolchover, Natalie. "A New Physics Theory of Life." *Quanta Magazine*, January 2014. https://www.quantamagazine.org/a-new-thermodynamics-theory-of-the-origin-of-life-20140122/.

Wolfram, Stephen. *A New Kind of Science*. Champaign, IL: Wolfram Media, Inc., 2002.

Wolpert, David H. "Physical Limits of Inference." *Physica* 237, no. 9 (2008): pp. 1257–1281. doi:10.1016/j.physd.2008.03.040.

Wright, Sewall. "The Relative Importance of Heredity and Environment in Determining the Piebald Pattern of Guinea-Pigs." *Proceedings of the National Academy of Sciences of the United States of America* 6, no. 6 (1920): 320–332. doi:10.1073/pnas.6.6.320.

Zimmer, Carl. "Hunting Fossil Viruses in Human DNA." *New York Times*, January 2010. https://www.nytimes.com/2010/01/12/science/12paleo.html.

INDEX